Canadian
AVIATION

Suzanne Kearns, Ph. D.

University of Western Ontario

Kendall Hunt
publishing company
4050 Westmark Drive • P O Box 1840 • Dubuque IA 52004-1840

Cover image © flag: Image copyright Micha Rosenwirth, 2009, under license from Shutterstock, Inc.

plane: D-JET003c Cover photo courtesy of Diamond Aircraft Industries

Kendall Hunt
publishing company

www.kendallhunt.com
Send all inquiries to:
4050 Westmark Drive
Dubuque, IA 52004-1840

Printed in the United States of America
10 9 8 7 6 5 4 3 2

This work is dedicated to my family:

Mike, Katie, & Sam
Eric & Dale
John & Mary Anne

Contents

Acknowledgments

A long list of individuals and organizations contributed to this text. The author wishes to express sincere thanks to the following individuals who contributed their writing, time, or support in the creation of this book.

Contributing Authors

The author is especially thankful to Jonathan F. Vance, from the History department at the University of Western Ontario, who contributed his knowledge and expertise to this book in Chapter 1—the History of Aviation in Canada.

Student Authors

Simon John Levesque

Jason Fogg

Joshua Norman

Ryan Clarke-Boles

Rupert Schuld

Colin Brown

Kenneth Polonenko

Consultations

Joan Finegan—UWO

Mitch Rothstein—UWO

Peter Rowntree—TSB

Don Enns—TSB

Chris Krepski—TSB

Brian Morris—Air Canada Jazz

Dominic Totino—Seneca College

Gary Anderson—Seneca College/Air Canada

Alex Lau—UWO Work Study Student

Wayne Gouveia—Air Transportation Association of Canada

Scott McFadzean—Diamond Aircraft

Captain Paul J. Teremchuk—Canadian Forces Recruiting Centre

Ron Singer—NAV CANADA

John Urban—NAV CANADA

Daniel Gooch—Canadian Airports Council

Greg MacDougall—CATSA

Bill Newman—Greater Toronto Airport Authority

Dan Cornell

Fred Jones—Helicopter Association of Canada

Kevin Psutka—Canadian Owners and Pilot's Association

Bill Welsh—Diamond Flight Centre

Jim Dow—Transport Canada

Kathie Keeley—Transport Canada (CARAC)

Mike, Katie, & Sam Kearns

Mary Anne & John Kearns

Lynn and Chris Murray

Eric and Dale Robinson

Mike and Amy Robinson

CHAPTER **1**

THE HISTORY OF AVIATION IN CANADA

OBJECTIVES

After you have read this chapter, you will:

- Understand the process by which humans came to achieve artificial flight

- Appreciate the impact of war on civil aviation in Canada

- See how Canada's place in international aviation changed over the twentieth century

- Understand the importance of lighter-than-air flight in Canadian aviation history

- Grasp the critical role that the federal government has played in Canadian aviation since 1909

As long as humans have walked the earth, they have probably longed to rise above it. Religion, myth, and legend are filled with tales of creatures endowed with the power to ascend into the heavens. Mohammed's flight from Mecca to Jerusalem and back on a fanciful flying creature, Pegasus the winged horse, angels and faeries, the winged bulls of Assyria, the legendary flying carpet, the winged gods Mercury and Hermes—all bear witness to the human desire to break the constraints of earthly existence.

The Beginnings of Flight

The transition from fancy to reality, however, was not so simple, for it was not clear how an ordinary mortal could be lifted into the air. The mythical Daedalus and Icarus attempted to imitate the birds by crafting wings of feathers and affixing them to their backs with wax, a solution that anticipated one of the most enduring assumptions about the best way to achieve human flight, the ornithopter concept. For centuries, monks, nobles, scientists and artisans alike believed that the best way to achieve artificial flight was to imitate the birds, using human strength to operate mechanical wings. Leonardo da Vinci was the most sophisticated of these thinkers, filling page after page with intricate sketches of avian anatomy and flying machines patterned after birds. Da Vinci wisely declined to test his ornithopters himself, but other budding aeronauts were less reluctant. Indeed, the annals of early flight are filled with intrepid pioneers plummeting to their deaths in primitive ornithopters or jumping from towers wearing cloaks stiffened into wings.

Roger Bacon, the thirteenth-century English scientist and monk, knew of these failed attempts, and was convinced that the obstacle was weight: the devices were too heavy to rise into the air using human strength alone. He turned his mind to a lighter-than-air flying machine, writing of thin copper globes, filled with "thin air or liquid fire," that would rise into the air and float in the atmosphere. Bacon may have been first to conceptualize the hot-air balloon in any detail, and as the centuries passed, the lighter-than-air principle seemed to offer the best prospects for success. In 1670, Francisco de Lana envisioned a small boat that would be carried aloft by four copper spheres emptied of air. He reasoned that if air had weight, the absence of air must have no weight; therefore, the globes would rise. He and Bacon were on the right track in assuming that gases had different weights and, therefore, different lifting properties. However, it would be another hundred years before this notion was successfully put into practice by the French Montgolfier brothers, who sent the first human aloft in 1783.

By the mid-nineteenth century, the aeronaut's vehicle had become much simpler than the Montgolfiers's ornately decorated craft. A small basket was suspended beneath a large bag made of canvas or silk, held within a mesh of rope. Lift was usually provided by coal gas, which was cheap and easy to produce, and the inflation process could take many hours, depending on the ambient temperature and other climatic factors. Once aloft, altitude could be controlled by dropping ballast (to ascend) or releasing gas (to descend). Beyond that, a balloon was completely at the mercy of the wind, which might carry both aeronaut and vehicle into trees, buildings, or a nearby lake. In time, the round balloon was replaced by a sausage-shaped gas bag beneath which was suspended a triangular wooden frame with a propeller at one end, a small internal combustion engine in the middle, and a rudder at the other end. The pilot sat astride the frame and controlled the rudder by means of a long cord. Because it could be guided, the craft became known as a *dirigible* (from the French *diriger*, to steer). The aeronaut also had to operate a hand pump to keep oil circulating into the engine and monitor the pressure of the air bag by periodically tapping it with his finger. His other primary task was keeping the airship level, which he did by moving back and forth along the frame.

The Air Age Comes to Canada

10 August 1840   *Eager New Brunswickers started to gather around lunchtime, and by late afternoon most of the vantage points around Saint John's Barrack Square were thronged with spectators. That day, Louis Anselm Lauriat of Boston would ascend in his balloon Star of the East. The weather was very clear with a gentle breeze from the southwest, and there was a buzz of excitement as Lauriat made his final preparations. Just after five o'clock, he commanded his assistants to cast off the ropes, and the Star of the East rose, slowly and tentatively at first, but gradually with more assurance. From the basket suspended beneath his balloon, Lauriat waved his hat and bowed gravely to the people below. They watched for nearly an hour as the balloon drifted north, then wafted to the east as the wind pushed it away from the city. As he drifted over the farmlands east of Saint John, Lauriat ascended farther, reaching a height of some 7300 feet. After nearly ninety minutes in the air, he brought the Star of the East to a safe landing in a field twenty-one miles from the city. He packed up his balloon, collected his meagre earnings, and returned to Boston by steamer.*

Balloonists like Lauriat gathered tidy sums doing display flights across North America. But the dream of heavier-than-air flight had not died. In Dayton, Ohio, Orville and Wilbur Wright had first developed an interest in flying as children, when their father bought them a small helicopter toy. In the late summer of 1896, the Wrights began to take notice of aviation news items that appeared in local papers. They read everything they could find on the subject, putting other people's theories to the test and, when they were found wanting, returning to the essentials of flight for inspiration. In an astonishingly short period of time they were building and testing gliders at Kitty Hawk, North Carolina, chosen because of its favourable climatic conditions. Better-than-expected flights in 1902 convinced them that it was time to take their next step: the addition of an engine to their glider. In September 1903, they crated up their new *Wright Flyer* and set out for Kitty Hawk. On 17 December 1903, Orville coaxed the *Flyer* into the air for a grand total of twelve seconds. The dream had finally been realized—powered flight was a reality.

Alexander Graham Bell, for whom the telephone was only one in a long list of successful inventions, had also been following the successes and failures of aviation pioneers, and was determined to bring powered flight to Canada. In Baddeck, Nova Scotia, the Bell estate became a magnet for the small but growing community of aviation enthusiasts. Among this group were four young men, energetic and technically minded, whom Bell hoped would help him achieve his dream of artificial flight. Casey Baldwin and Doug McCurdy had both attended the University of Toronto. American Tom Selfridge, an artilleryman from California, had met Bell in Washington and received an invitation to Baddeck to observe the experiments for the U.S. Army. The fourth man, Glenn Curtiss, from Hammondsport, New York, was a mechanical genius who owned a chain of bicycle stores and manufactured light and powerful aero engines. Bell and his four young associates worked independently of one another at Baddeck for most of the summer of 1907, until Mabel Bell offered $35,000 to formalize their partnership and cover expenses. On 1 October 1907, the Aerial Experiment Association [AEA] came into being "for the purpose of carrying on experiments relating to aerial locomotion with the special object of constructing a successful aerodrome." The five would first attempt to complete Bell's pet project, a giant human-carrying kite. In just two months, that goal was attained. On 6 December 1907, the *Cygnet*, carrying Tom Selfridge, was pulled into the air by a small steamboat and remained aloft for a full seven minutes.

Baldwin, Curtiss, McCurdy, and Selfridge then relocated to Curtiss's motorcycle factory in Hammondsport to work on airplanes. Over the next few months, the AEA recorded a string of successful flights, first Selfridge's *Red Wing*, and then Baldwin's *White Wing* and Curtiss's *June Bug*. Before McCurdy's *Silver Dart* was completed, tragedy struck. In August 1908, Selfridge was invited to accompany Orville Wright on a number of demonstration flights, but on 17 September 1908 at Fort Myer, Virginia, Wright's aircraft crashed; Orville was seriously injured, and Selfridge became the first fatality in an airplane crash. However, Selfridge's death did not deter his associates. Mabel Bell offered another $10,000 to extend the AEA for a further six months so that its work could continue over the winter. On 23 February 1909, the *Silver Dart* was wheeled from its shed onto the frozen lake. A good number of people had gathered, and, as they waited patiently, McCurdy opened the throttle and the aircraft began to slide along the ice. It gathered speed for about ninety feet and then slipped into the air. McCurdy took the aircraft up to forty feet, flew for about half a mile, and lightly landed the *Silver*

F.H. Hitchins Collection.

Figure 1.1 *A First World War postcard that emphasizes the role of the airplane as a means of observation.*

Dart back on the ice. It was all over in ninety seconds, perhaps before many of the spectators realized that they had been watching history in the making.

And then . . . nothing. Flying clubs sprang up in Winnipeg, Toronto, and Montreal, but they were tiny and failed to attract many people who had actually flown. Canada's first international air show was held at Montreal's Lakeside Park in 1910; many others followed, but they sometimes featured more crashing than flying. The Canadian government, to whom McCurdy proposed the idea of an airplane with military utility, saw it as nothing more than a trifle and declined to pursue the idea. Five years after the first flight of the *Silver Dart*, there were still only a handful of Canadian pilots (although a steady stream of American aviators came north to ply their trade), and the aero clubs established with such confidence after that first flight had quietly collapsed. Doug McCurdy spent much of his time after 1909 in the United States, where aviation found a more congenial atmosphere, while Casey Baldwin stopped flying in 1911, to turn his skills to other engineering projects. Bell continued to speak publicly in support of flying, but wound up his aviation experiments in 1912. The Canadian Aerodrome Company that the three of them had founded with such high hopes withered through lack of interest. If the technology was to progress, a catalyst would be needed.

The First World War

The catalyst was war. In the summer of 1914, the European powers drifted towards war, pulling much of the world with them, and all sides were soon looking for an advantage on the battlefield. Official interest in the airplane quickly revived, but the first weeks of the war suggested that the military had been right in the first place. The first British reconnaissance mission, on 19 August 1914, was a disappointment, both pilots getting lost without locating the units they were sent to find. Days later, a German reconnaissance pilot reported a mob of panicked and disorganized British soldiers, but his seemingly demoralized enemy had been engaged in a soccer game. A Canadian Aviation Corps was conjured out of thin air in September 1914, with a strength of two officers and an American-made Burgess biplane that even its manufacturer admitted was in rough shape. The Burgess was damaged in transit and never flew again. By the time Lieutenant W.F.N. Sharpe of Prescott, Ontario, was killed in a training crash in February 1915, his commander, Captain E.L. Janney of Galt, Ontario, had already been discharged and the Canadian Aviation Corps had ceased to exist.

But the airplane was coming into its own as the war of movement ended in the fall of 1914. As the opposing armies dug into the line of trenches that eventually stretched from Switzerland to the English Channel, reconnaissance, pinpointing everything from supply dumps to machine-gun posts, remained vital. At first, the armies relied on hastily trained observers to sketch or describe what they saw as they flew over enemy lines, but aerial photography soon supplemented their work. Over the course of the war, the British took half a million aerial photos, which they could compare over time to discern changes in the enemy's defensive positions. Just as important was

We have eyes for observation,
When far up in the blue,
And in the Fight for Home and Right
We've hands for action too;
O'er Land and Sea you'll find us
To win we steer our course,
To duty loyal, by foemen feared—
The Empire's Royal Air Force.

FROM ONE OF THE "ROYAL AIR FORCE."

Collection of Jonathan F. Vance.

Figure 1.2 *A First World War postcard that emphasizes the role of the airplane as a means of observation.*

artillery spotting, both locating enemy batteries and directing the fall of shot on targets, a complicated technical exercise that used wireless transmitters or signalling lamps to convey locations or map references.

These tasks became so critical that those aircraft became targets of enemy scouts, and had to be accompanied by friendly scouts as escorts. By 1917, large swarms of sixty to seventy aircraft could be seen engaged in battle over the Western Front on any given day as each side attempted to gain command of the air. The demands for pilots and airplanes grew enormously, and Canada was able to do its part. After the success of the AEA, Glenn Curtiss had expanded his Hammondsport shop into the largest aviation company in North America, and he saw the war as the perfect opportunity to return to Canada, with McCurdy as his local agent. Again, McCurdy tried to sell the Canadian government on an air force, offering to manufacture the airplanes at a Curtiss factory in Toronto and train one pilot for every machine the government bought. Ottawa balked, but the British Royal Flying Corps accepted the offer. They would purchase training aircraft in the United States and send them to Toronto for modification; they would also recruit pilots in Canada and train them at Curtiss flying schools across the country. The agreement lasted for two years and produced over 250 pilots, but it was far too little to satisfy the demand for air crew. In December 1916, the British corps was replaced by RFC Canada, which would establish twenty training bases across Canada. At the same time, the British government bought the Toronto Curtiss factory, renamed it *Canadian Aeroplanes Limited*, and started to manufacture the Canuck, a Canadian variant of the enormously successful Curtiss JN4 Jenny. By the end of the war, the factory had built over 1200 Canucks, as well as fifty Felixstowe flying boats.

The Felixstowes were designed for maritime patrol duties, but their range and carrying capacity raised an interesting possibility: why not bypass the battlefield altogether? The new technology could carry the war right into the enemy's capitals and rain bombs on Berlin, London, or Paris until the enemy capitulated. The first bombs dropped on British soil fell on Christmas Eve 1914, but did no damage; on 10 January 1915, the German airship service was given permission to begin a limited aerial bombardment of England. The first raid came nine days later, killing four civilians and injuring sixteen, and as the months passed the night raiders came with increasing regularity. The vulnerability of airships to attack and their inability to bomb accurately pushed the Germans to develop larger attack aircraft, the multi-engine Gotha and Giant bombers that continued combing British targets until the end of the war. The raids had next to no practical impact on the war, but their psychological impact was immense.

Terror-Bombing

In 1915, for the first time in history, civilians were exposed to aerial attack. In the hundreds of raids, the German airships (known as Zeppelins, the trade name used by one manufacturer) and airplanes mounted against Britain over the next three years, casualties were low but the indiscriminate nature of the attacks horrified the world. On 31 May 1915, bombs fell in east London. Only seven people were killed, but four of them were children, including a little girl and her baby sister, burned to death when incendiary bombs hit their home in Stoke Newington. The first Gotha raid came on 25 May 1917 at Folkstone, the main embarkation port for British troops heading to France. Two weeks later more than 160 Londoners were killed, including fifteen children who died when the Upper North Street School in Poplar was devastated by a bomb. In January 1918, one of the Giants dropped a single bomb in front of the Odhams Printing Works in central London. The explosion collapsed a wall, tipping the printing presses into the basement, where more than 500 people had taken shelter, and igniting rolls of newsprint. The resultant inferno killed thirty-eight people and injured eighty-five. In a war that could see ten thousand soldiers killed in a single day, the deaths were a trifle, but they sowed in the public mind an intense fear of terror-bombing.

By war's end, over 20,000 Canadians had served in the RFC and the Royal Air Force (as it was known after April 1918), and as much as 40 percent of the RFC's pilots on the Western Front (and 25 percent of its total strength) were Canadian. Canada produced 127 aces (pilots with five or more aerial victories), and together they shot down some 1500 enemy aircraft. Aviators such as Billy Bishop and William Barker became household names who could draw huge crowds to their public appearances. What would all of those skilled pilots, air crew, and mechanics do once peace had returned?

ZEPPS!!

Figure 1.3 *This 1916 postcard offers a humorous look at the public's growing fear of air raids.*

The Interwar Era

For air-minded Canadians, the end of the war in November 1918 brought a golden opportunity. The airplane had demonstrated its utility during the war—they had to prove that it could be just as valuable during peacetime. One thing was certain: there would be plenty of airplanes to go around. Soon after the war ended, the government decided to dispose of the aircraft used to train pilots in Canada. In the spring of 1919, a Toronto entrepreneur was advertising Canucks at $2000 apiece. Soon, there were other kinds of aircraft available, fostering the growth of a small aviation manufacturing sector in Canada, and an even healthier import trade. The Curtiss H2SL flying boats, bought as war surplus and pressed into service to become Canada's first bushplane, were replaced by the Vickers Viking, manufactured by Canadian Vickers in Montreal from a British design, and the much more successful Vickers Vedette, the first airplane to be built specifically with Canadian conditions in mind. The American manufacturer Fairchild had such good sales in Canada with its FC1 and FC2 monoplanes that it established a factory in Montreal to manufacture them. The Dutch company Fokker and the German company Junkers both manufactured excellent aircraft that were well suited to operating in the Canadian wilderness.

But where would the money for gas and oil, spare parts, unexpected repairs, and a little profit come from? With few stable sources of revenue, many pilots turned to the one pursuit with any hope of remuneration: entertaining the public. And so was born the aerial barnstormer, taking the nickname from the itinerant showmen who crossed rural America, performing in barns, town halls, or any other venue that would have them. Typically, the aerial variant buzzed a small town to draw the people into the streets, sometimes dropping handbills to advertise their services. He landed in a nearby farmer's field and offered rides, for a flat fee or a certain sum per minute or per pound of the passenger's weight. Once the interest and disposable income of that town was exhausted, he would move on to another.

The enterprising aviator could also take advantage of any public gathering to earn a few dollars. Depending on the season, pilots might travel the circuit of country fairs, bidding for contracts with fair organizers and performing aerobatics for the crowds. Within a year of the war's end, every fair worth its salt featured an aerobatic display. Aviators also found that they could pick up extra income by renting out their aircraft for sales gimmicks. Stan McClelland in Saskatoon turned his Curtiss into an aerial billboard and plastered it with advertisements, while Russell Smith of Ottawa started the Aerial Services Company to "bomb" every city, town, and village with advertising literature. In May 1919, Roland Groome delivered a special edition of the Regina *Leader* from the Saskatchewan capital to Moose Jaw, Indian Head, and Qu'appelle. When Ernest Hoy made a pioneering flight over the Rockies from Vancouver to Calgary in August 1919, he carried with him sweaters from Universal Knitting in Vancouver to be displayed in the window of the Diamond Clothing Company in Calgary.

But for the pilot who found these pursuits too tame, there was a far bigger challenge: crossing the Atlantic. In 1913, British newspaper magnate and aviation enthusiast Lord Northcliffe had announced a prize of £10,000 to the first person to fly between North America and Great Britain or Ireland, but the coming of war had temporarily halted those attempts. However, the war had also fostered technological developments that brought the flight within reach. The multi-engine bombers intended for British reprisal raids against Berlin had the range to cross the ocean, as did the big flying boats built to protect the coasts of North America and Britain. Northcliffe announced in July 1918 that the prize money was still up for grabs, and when the British government released Royal Air Force officers to enter the fray, contenders for the prize came in droves.

In the summer of 1919, the contestants started to gather in Newfoundland. Most of their attempts ended in either embarrassment or tragedy. Even the U.S. Navy, with an unlimited budget, could manage to get only one flying boat across the Atlantic; three others and a dirigible failed in the attempt, and the successful flight was made possible only with a chain of warships to act as floating beacons and a stopover in the Azores. The best hope, the Vickers Vimy *Atlantic*, a two-engine bomber that had been developed too late in the war to see action, was ready for its first flight on 9 June, but not until 14 June did pilot John Alcock decide it was safe to fly. He and navigator Arthur Whitten Brown packed the *Atlantic* with some food, coffee, and letters, topped up the tanks, and trundled off down the field. The Vimy struggled for height, and then disappeared below the hills. The crew was sure it had crashed, but seconds later a cheer went up as the Vimy reappeared, ascending slowly towards the east.

Figure 1.4 *One of the contestants for the trans-Atlantic flight arrives in Newfoundland in 1919—still in its packing crate.*

Brown would later recall the flight as his single most terrifying experience. Soon after leaving Newfoundland, one of the engine exhausts burned off, so the roar was deafening. Then there was a dizzying spin out of a cloud bank, from which Alcock recovered with barely a hundred feet to spare. To avoid more thunderheads, Alcock climbed higher, but the control surfaces, engine intakes, and instruments began to ice up. Finally, at 8:15 A.M., Alcock peered through the gloom and thought he could make out some islands. Soon a larger land mass came into view, and ten minutes later the Vimy crossed the Irish coast. They circled the town of Clifden and gently put the aircraft down on what they discovered too late was a peat bog. The wheels dug in and the stalwart Vimy buried its nose in the peat. But Alcock and Brown had bridged the Atlantic in a single hop.

Over the next eight years, other crews would repeat their achievement, but not as many as one might imagine. Aircraft remained primitive, and most were no match for the brutal weather conditions over the North Atlantic. More aircraft were lost in the attempt than made it across, and those that did make it carried more than one crew member. The solo flight across the Atlantic remained elusive. But in 1927, a handful of pilots again gathered on both sides of the ocean to make the attempt. As in 1919, accidents reduced the field, leaving an unlikely challenger: a little-known Nebraska air mail pilot named Charles Lindbergh.

On 20 May 1927, at 7:51 A.M., Lindbergh's Ryan monoplane, little more than a flying gas tank, splashed along a sodden Roosevelt Field in New York. With barely a thousand feet of grass left, he muscled the aircraft off the ground. It lurched over the telephone wires at the end of the field, struggled for height, and eventually headed out towards Long Island. Against all odds, on 21 May 1927, as a calm, clear night settled over Paris, the *Spirit of St. Louis* touched down at Le Bourget airport. The haggard Lindbergh clambered out of the cockpit to be greeted by a crowd nearly driven mad by excitement and anticipation. As many as 200,000 people jammed the airport, and millions more followed his every word in the world's newspapers over the next few weeks. He was, quite simply, the most famous man in the world in 1927. The soft-spoken mid-westerner had almost become divine. People referred to him as an angel or a god-man, and politicians, generals, and captains of industry clamoured to spend a few minutes in his presence. Charles Lindbergh hadn't just flown across the Atlantic; he had captured the attention of the world.

Hitchins Collection.

Figure 1.5 *Charles Lindbergh (right, meeting his adoring public, 1927).*

But ironically, Lindbergh's achievement was of limited use to the air lobby. They needed to demonstrate, not that flying was a miracle accomplished by human gods, but that flying should be a normal part of daily life. And since the end of the war, they had been carefully cultivating public opinion to that end. The Aero Club of Canada, chartered in 1917, scored its first big coup by lobbying the federal government to create the Air Board in June 1919, under controller of aviation John A. Wilson, and establish the first air regulations to cover the operation of aircraft. The city of Regina grabbed most of the firsts, having the first licensed airport, the first licensed pilot (Roland Groome), and the first registered aircraft (G-CAAA, of Groome's Aerial Service Company).

A decade later, it was the Canadian Flying Clubs Association, established in 1929, that emerged as the most aggressive aviation lobby in the country. Since the early 1920s, the air lobby had been urging the federal government to provide more support for civil aviation, and on 23 September 1927, Ottawa finally announced a program of government assistance to airplane clubs: the government would pay $100 towards the training costs of each pilot who qualified for a license; the student would cover the other $150. The upshot, as the air lobby had hoped, was an explosion in the number of active flying clubs across the country. On 1 May 1928, the Toronto Flying Club became the first in Canada; it was followed the same month by clubs in Saskatoon, Montreal, Hamilton, Winnipeg, and Regina. By the end of 1928, there were sixteen clubs with 2400 members. In the Empire, only Britain could boast more light airplane clubs than Canada.

The CFCA became the driving force behind the air show boom of the 1930s. From the small-scale air shows mounted by the members of the local flying club to the huge national air tours, this was aviation's travelling road show, taking airplanes and pilots to eager crowds in villages, towns, and cities. In Alberta, the big draw was the air circus. Local pilots would organize eight or ten aircraft and fly from town to town putting on a show that included all the standard tricks—parachute jumps, formation flying, stunting, air races, and penny-a-pound rides. A typical gag was a pilot dressed up as an elderly woman who climbed into the aircraft for a closer look; of course, the airplane would then "accidentally" take off, much to the delight of the crowd. Slightly more sophisticated were the regional air tours that became popular in the early 1930s, such as the Manitoba Goodwill Air Tour, first held in 1931. The tour was such a success that it was repeated the following year and extended to two weeks so more towns could enjoy the show. Aviation boosters in Atlantic Canada got together to organize the first Maritime Air Tour in August 1932. The tour was followed by the gala Maritime Air Pageant in Charlottetown, which the province's air-minded lieutenant-governor recommended "not only as a means of entertainment, but as a practical demonstration of the value of the airplane as a safe means of transportation."

These immensely popular regional air tours were themselves dwarfed by the Trans-Canada Air Pageant; it was modelled on the National Air Tour, which toured the United States in the 1920s and 1930s and usually included a few Canadian stops. The Pageant was equal parts entertainment, education, and marketing. Aircraft manufacturers had a significant presence, showcasing their newest products with demonstrations, fly-pasts, and ground displays. There were segments that highlighted the speed and load-carrying capacity of aircraft, and examples of aerobatics and formation flying. To lighten the mood, there was always a novelty act, frequently some variation on the senior-citizen-accidentally-flies-an-airplane routine. By all accounts, the Trans-Canada Air Pageant was a hit. The CFCA, which organized the event, estimated that the first tour drew more than 300,000 spectators in the twelve cities it visited; the crowds in subsequent years were even larger. This success is even more remarkable when it is noted that the air lobby was a tiny group—in 1930, there were still only 539 licensed pilots in Canada—but they were remarkably successful in spreading the message.

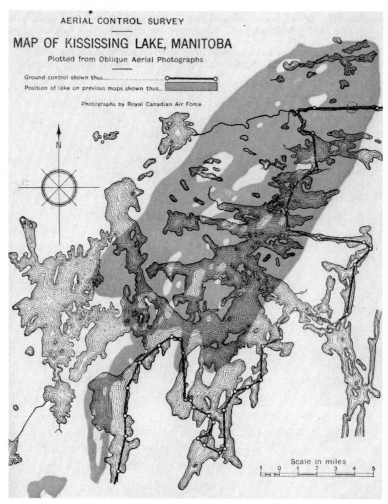

AERIAL CONTROL SURVEY

MAP OF KISSISSING LAKE, MANITOBA

Plotted from Oblique Aerial Photographs

Ground control shown thus.......................
Position of lake on previous maps shown thus...

Photographs by Royal Canadian Air Force

Scale in miles

Department of National Defense, *Report on Civil Aviation for the Year 1923* [Ottawa: King's Printer, 1924].

Figure 1.6 *The miracle of aerial surveying. The dark, arrowhead shape is how Lake Kississing appeared on earlier maps.*

Central to that message was the notion that aviation could transform the economic life of Canada, and ultimately the nation itself. Within a few years of the war's end, the airplane was already showing that techniques of aerial photography, so effective during the war, could be just as useful in peacetime. The Department of the Interior mounted a very successful test survey, after which the Air Board and the Canadian Air Force began more ambitious operations. The first two years of surveys, in 1922–23 and 1923–24, were strictly experimental, carried out during the regular patrols by the CAF. In succeeding years the amount of territory covered by aerial survey flights grew dramatically: in 1924, the flights covered some 40,000 square miles, 70,000 square miles in 1929, and a record 108,000 square miles during the 1935–36 fiscal year.

But even before the aerial survey was mounted, the private sector (beginning with the forestry industry) had grasped the value of the airplane in resource management. Perhaps the first organization was the St. Maurice Forest Protective Association, a consortium of Quebec lumber companies that had purchased a number of war-surplus flying boats. In June 1919, ex-war pilot Stuart Graham made the first timber survey flight over the Lac à la Tortue region of Quebec and at the end of the season, the St. Maurice forest manager declared that ten men working on foot could cover in a month only a quarter of the territory recorded in one day of aerial photography flights. An experienced "timber cruiser" could classify varieties of trees based on their colour and shape, and determine both the density of timber and the ease with which it could be logged. Access routes could be mapped out, obstacles noted, burned over or swampy areas marked—all in one survey flight. The forestry industry lost no time in harnessing the potential of the aircraft to maximize profit.

Price Brothers was one of the first companies to establish a flying branch, and began air patrols over its holdings in the Lac-Saint-Jean area of Quebec in July 1920. Laurentide Air Service, established in 1921 by one of the partners in the St. Maurice consortium, emerged as the biggest player in the game, winning contracts from the Ontario government and private concerns to do forest reconnaissance flights from its base at Lac-à-la-Tortue. The company doubled its air fleet between 1922 and 1923 and established a larger headquarters at Trois-Rivières. Though it ceased operations in 1925, Laurentide was one of the great success stories of postwar aviation in Canada.

But cataloguing forest resources was only part of the job; protecting them was just as important, and more challenging. The greatest threats to Canada's forests, then as now, were parasites and fire. The sawfly or hemlock looper could devastate timber stands; the federal Department of Agriculture estimated that a spruce budworm infestation in eastern Canada in the early 1920s destroyed as much as 150 million cords of pulpwood. Fire could be even more disastrous. Manned watchtowers covered only a tiny portion of Canada's forests, and a fire could burn for days or weeks before it was detected. Quite apart from the potential loss of life, the loss of valuable timber was immense, averaging 900,000 acres a year in the five years up to 1924 and peaking at 2.2 million acres in 1923. Faced with such losses, forest companies were willing to try almost anything to safeguard their assets.

Determining the spread of parasite infestation by air was simple: an aerial survey of a spruce budworm outbreak near Lake Abitibi in 1920 yielded "more information from the air in a day as to the extent of the outbreak than they could in a season from the ground," John Wilson reported. On the west coast, aircraft from the RCAF's Jericho Beach flying boat station trapped white pine blister rust spores at different altitudes, so scientists could determine how the disease travelled to infect new areas. Before long, governments and forestry companies discovered that outbreaks could be fought, as well as monitored, from the air. The techniques for fighting parasite infestation by air were relatively complicated and took some time to perfect, but the first aerial forest dusting occurred from 18 June to 27 July 1927 in Cape Breton. By the end of the 1930s, government and forestry officials had a whole new arsenal with which to fight parasite infestations.

The possibility of fighting forest fires from the air had been broached in 1917, and the forestry companies quickly realized that their survey flights could double as forest fire patrols. The St. Maurice Forest Protective Association made forty flights in 1919, and the following year, government air stations were established at Jericho Beach, British Columbia; Morley, Alberta; and Roberval, Quebec, to provide forest fire patrols. Ontario went a step farther. In the early 1920s, it had contracted Laurentide to fly forest patrols, but, in 1924, the cabinet decided to establish its own flying branch. And so was born the Ontario Provincial Air Service, initially with thirteen flying boats purchased from Laurentide. Headed by Roy Maxwell, a swashbuckler with a blue Cadillac roadster and a weakness for high leather boots, the OPAS established its main headquarters at Sault Ste. Marie, and for a few years was the largest aviation organization in Canada. In 1930, its thirty-three aircraft and seventy-nine flying personnel completed nearly 12,000 flights over 875,000 miles and achieved its primary goal of protecting the province's forests. By 1927, its aircraft were patrolling 125,000,000 acres of forest; losses that year were only 35,000 acres. The government estimated that 90 percent of the province's forest fires never spread beyond 100 acres because of the OPAS's ability to deliver fire-fighting crews quickly and efficiently.

Mineral resources, too, could be discovered and developed by air. In July 1920, Stuart Graham flew a prospector into a remote area of Quebec to stake the first mining claim by air. In 1921, geologist E.L. Bruce flew from Sioux Lookout to Lake St. Joseph in northern Ontario in a little over an hour, a distance that had taken him four days by canoe the previous summer, and was persuaded that the airplane could save the prospector or geologist much time and effort. It could never replace the time-consuming practice of tramping through the bush with a hammer to gather samples, but it certainly made getting in and out much easier. A prospecting party, instead of spending weeks canoeing and hiking, could be flown to any lake or river and begin work immediately. If the area proved a disappointment, they could easily be relocated; if prospects were good, the party could be supplied by air for the entire season and brought out in the fall. Moreover, a single aircraft could easily supply half a dozen prospecting parties.

The mining boom was just what many struggling pilots and aviation companies were looking for. Laurentide, still reeling from the loss of its Ontario government contracts, filled its balance-sheet void by opening the first successful regularly scheduled service in Canada, to the gold fields of Rouyn, Quebec. Hundreds of miles to the west, an even bigger drama was about to be played out. On 25 July 1925, a party of four prospectors struck gold near the Manitoba–Ontario border. The Red Lake gold rush, the last classic gold rush of the twentieth century and the first to be serviced by air, was on. Aviation boosters were fond of noting that the Red Lake and Rouyn runs were the world's only self-sustaining, regularly scheduled air routes.

An Aerial Gold Rush

Jack Hammell, the Toronto mining promoter to whom the four prospectors looked for financing, knew that he had to begin a surface exploration immediately, before the freeze-up and before word got out and eager grubstakers flooded in. But there was not time to get a party in by foot. The only solution was the airplane. Hammell chartered five flying boats from the OPAS, taking in fifteen tons of supplies and seven workers to assess the site. Hammell's haste was justified, because by early 1926 hopeful prospectors were coming to Red Lake from across North America. Mostly they came by dogsled, snowshoe, and canoe, but Jack Elliott, a First World War pilot with a fledgling flying company, sensed there was money to be made. He crated up a couple of Jennys fitted with skis, sent them by rail to Hudson, Ontario, and was soon charging a dollar a pound to fly people in and out of Red Lake. At the same time, Doc Oaks and Tommy Thompson resigned from the OPAS to establish Patricias Airways Exploration Limited. It took nearly four weeks to get their brand-new Curtiss Lark from the factory in New Jersey to Red Lake, and through 1926 the two air services operated at capacity supplying the mining concerns. For a brief point in time, Red Lake laid claim to being the busiest airport in the world, in the number of take-offs and landings on a daily basis.

But more important, the mining boom was the catalyst for the creation of Canada's first great airline, Western Canada Airways. James Richardson was a young, aggressive, and visionary financier who was always on the lookout for new opportunities and became convinced that mining and aviation could be a profitable mix. On 10 December 1926, WCA came into being, with Richardson as president and Doc Oaks, who had resigned from Patricia Airways, as manager. Five years after its foundation, it had become a complex network of air routes served by seventy-two aircraft—more than the federal government owned—that flew nearly 1.9 million miles and carried over 1.1 million pounds of freight. Its operations were varied—passenger services, fisheries patrols, mercy flights, timber-cruising, rescue missions, pilot instruction, charter services—but its staple was serving the prospecting and mining sectors. In short order, WCA would grow from a modest operation serving the mining districts of northern Manitoba and Ontario, into the leading air freight company in the world.

Another of WCA's revenue staples was flying the mail, but that was one of the last services to get off the ground in Canada. The United States Postal Service began experimenting with air mail in 1917, and the following year instituted a service linking New York and San Francisco. Its delivery rate of over ninety percent inspired others to recommend a similar experiment for Canada, but despite a successful cross-Canada mail flight in October 1920, the Post Office Department would give permission for only the Aero Club of Canada to mount a number of experimental airmail flights between Toronto and Ottawa, using its own resources—and its own stamps. The Post Office refused to believe that airmail was anything but a luxury. In the spring of 1920, faced with a suggestion for an experimental service between Moncton and Charlottetown, Postmaster-General R.M. Coulter replied that "he had never known of Prince Edward Island to be in any hurry for mail." When a service between Winnipeg Beach and Rice Lake, in eastern Manitoba, was mentioned, Coulter observed that only the mining companies were interested, and if they were that keen, they should foot the bill. Not only were postal officials put off by the potential costs and the projected 8.5 percent loss rate (which they regarded as unacceptably high), but they also pointed out that most of the interest had been shown, not by the general public clamouring for speedier mail delivery, but by aviation interests looking for employment. The Post Office Department would give official sanction to airmail runs by private companies, but declined either to initiate its own services or bear the costs or responsibility for the privately-run services.

Not until 1927 did the Post Office change its policy, when it let a contract for an airmail service to pick up mail from trans-Atlantic liners off Rimouski and fly it to Montreal at, it was hoped, a considerable saving in time. However, the first attempt, on 9 September 1927, was not a success. The flying boat assigned to the run was damaged by heavy swells at Rimouski and crashed while attempting to take off. The 500 pounds of mail from the liner *Empress of France*, intended to reach Montreal a day before the ship docked, was salvaged from the wrecked airplane but not delivered until well after the *Empress* had reached Montreal. However, nine successful flights were made before the middle of November, both to incoming and outgoing liners, and in some cases the service shaved ninety-six hours from the delivery times of mail between Britain and Canada.

The Rimouski run was oriented towards business and official correspondence, but three other contracts awarded in 1927 focused directly on reducing the isolation of certain settlements. They linked Leamington, Ontario, with Pelee Island, in Lake Erie, the southernmost settlement in Canada; Murray Bay, Seven Islands, and Anticosti Island, in the Gulf of St. Lawrence; and Moncton and the Magdalen Islands. A fourth, connecting Rolling Portage with the Red Lake area in northern Ontario, was intended to serve the booming gold fields. Finally in 1928, after considerable lobbying, the Post Office agreed to expand airmail services between urban centres. Canadian Airways Limited[*] received the contract to operate the Montreal–Toronto route, and the following year a Toronto–Buffalo run was added. The Post Office also experimented with a number of other routes in 1929—Moncton to Quebec, Montreal to Moncton via Saint John, Montreal to Ottawa, Moncton to Charlottetown, and Montreal to Halifax—with enough success that all services, except the last, were maintained. There were further extensions in March 1930, when Western Canada Airways won the contracts for Winnipeg–Regina–Calgary and Regina–Edmonton services.

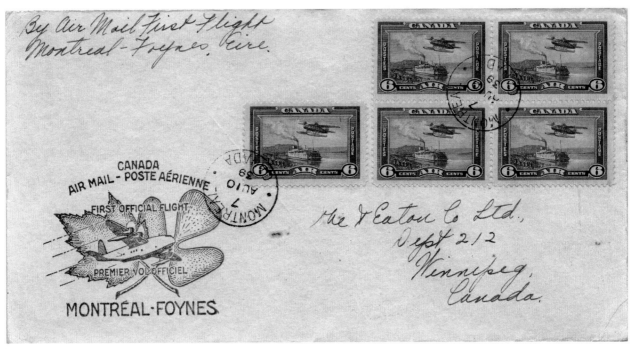

Collection of Jonathan F. Vance.

Figure 1.7 *A souvenir cover to mark the first airmail flight across the Atlantic, 1939.*

* Canadian Airways Limited, based in Montreal, was acquired by the parent company of Western Canada Airways in 1930, with the new entity adopting the name Canadian Airways Limited.

The Depression pushed many aviation companies to the brink of collapse and killed the airmail project. In January 1931, the Montreal–Quebec run was cancelled, and, in March 1931, the Winnipeg *Free Press* revealed that the new Conservative government of R.B. Bennett was contemplating deeper cuts to the airmail network. In February 1932, the government announced the first of many cancellations that would ultimately see the Canadian airmail system reduced to a handful of scattered intercity routes and some services to isolated communities and resource areas. The main intercity routes would not be reinstated until later in the decade.

But if the Depression spelled the end of many airmail routes, it also made possible the realization of another great national project: the trans-Canada airway, a chain of airfields stretching from coast to coast, the groundwork for a convenient, safe, modern system of air transport. Through the 1920s, Canada had only disconnected routes that sprang up as enterprising aviation companies established local services based on municipal airports, private landing fields, and bodies of water. Some major cities were without municipal airports, and large stretches of the country lacked even emergency landing fields. The trans-Canada airway was intended to solve all that.

According to the original plans, the trans-Canada airway would comprise nineteen main airports, twenty-four intermediate fields, and forty-two emergency fields. Municipal authorities would be responsible for the costs of their facilities, but the federal government would contribute the resources for the rest of the system: the intermediate and emergency fields, radio and meteorological services, and half the cost of improving lighting at municipal airports. The surveys began in 1928, with construction commencing the following year. The western section between Winnipeg and Edmonton was completed first, and by 1931 radio beam stations had been installed across the prairies. The Depression, which many feared would be the end of the airway, actually saved it, for it provided the federal government with a ready source of cheap labour. Between October 1932 and June 1936, airfield construction in British Columbia, Ontario, Quebec, and the Maritime provinces was done by unemployed men working under a federal relief scheme. When the government discontinued the scheme, the Civil Aviation Branch continued the work. By the end of 1938, there were modern airports or intermediate and emergency landing fields at least every 100 miles between Vancouver and Montreal.

It seemed self-evident that, now there was a national airway, there should also be a national airline, but how should that airline be created? One option was to use government subsidies to turn a regional carrier into a national one. Canadian Airways Limited seemed the natural choice, at least in the eyes of its very vocal supporters. It was the largest airline in the country, with a presence in every region. It had flown most of the airmail routes and employed some of Canada's most famous and experienced pilots. Its supporters had to admit that CAL always seemed on the brink of financial collapse, but they blamed government parsimony in awarding airmail contracts. In their eyes, no company was better suited to become Canada's national carrier. Unfortunately for Richardson and his allies, many people saw things differently. One postal official deplored "the slip-shod method of operation which has been so characteristic of this Company" and thought CAL should confine its operations to bush flying. An interdepartmental committee agreed, noting that CAL would be an inappropriate choice for a national airline because its organization for controlling scattered bush operations did not suit trans-continental services. The conclusion of many observers, one in which the federal government eventually concurred, was that a new corporation could best serve as Canada's national airline.

The negotiations to create that airline were protracted. Various possibilities, including full government ownership, joint ownership by Canada's railways, or the expansion of CAL after hiving off its bush operations into a separate company, were rejected. Eventually, on 22 March 1937 Minister of Transport C.D. Howe presented to the House of Commons the bill that would create Trans-Canada Airlines. The company's operations would be confined to carrying mail, express packages, and passengers on intercity routes; it would not compete with other carriers in bush flying. The new carrier took over its first route, the Vancouver–Seattle mail run, on 1 September 1937, and, on 1 February 1938, began flying between Winnipeg and Vancouver. The success of those flights convinced the Post Office to begin including mail shipments, again on an experimental basis; this service was regularized on 1 October 1938. On 1 April 1939, TCA began its daily service between Montreal and Vancouver. At last, Canadians could cross the nation by air.

But the cross-country flight was made up of a number of shorter legs; there was as yet no airplane capable of long-haul flights, particularly between North America and Europe. The flights of the Vickers Vimy and the *Spirit of St Louis* had demonstrated how far trans-Atlantic was from becoming a going concern through the interwar era, so the world's hopes rested with the dirigible. The C5 had been a dismal failure, but the embarrassment was soon forgotten, for in July 1919, a new British rigid airship, the R-34, cast off from its base in Scotland to make a nearly flawless flight to New York and back. The success seemed to offer proof that the airship, not the airplane, was the future of long-distance travel, and in May 1924, the British government announced that the Air Ministry would undertake a massive research and development program to build a brand new, five-million-cubic-foot airship, and mooring masts in England and India so that an airship service could link the British Empire. A private firm, the Airship Guarantee Company, would receive a contract to build a second airship to the same specifications.

Construction of both ships began in the summer of 1926, the R-100 at Howden in Yorkshire and the R-101 at Cardington. The process was not unlike assembling an immense child's toy. The skeleton of each airship consisted of huge rings of stainless steel and aluminum girders and tubes, joined by longitudinal ribs. The luxurious passenger compartment was located in the belly of the airship, surrounded by the immense gas bags. Made of cotton lined with goldbeater's skin, a thin membrane obtained from the intestines of cattle, the R-100's fifteen gas bags had a total area of more than ten acres. Once the frame had been assembled and the bags mounted, the entire airship was covered with strong linen, pre-doped to make it watertight. Although the two airships looked similar, there were significant differences. The R-100's design team relied on tried-and-true construction methods wherever possible, including standardized parts in the frame and variants of the gasoline engines that had powered British dirigibles since the First World War. The R-101 team chose revolutionary methods of design and construction to create the most advanced airship in the world. They were supremely confident in R-101's design, insisting it would be the strongest airship ever built.

As construction proceeded, empire politicians were trying to hammer out a co-ordinated air policy at the 1926 Imperial Conference in London. Sir Samuel Hoare, Britain's Secretary of State for Air, opened the deliberations on 28 October with a stirring summary of imperial aviation efforts and prospects for the future. Land and sea transportation, he observed, was unlikely to improve enough to make any appreciable difference in travelling time around the empire. But in the not too distant future, the farthest reaches of the empire would be days, not weeks, away from the mother country. Airplanes will be used on shorter routes, he believed, but "airships will carry out the long-distance non-stop air journeys of the future." Canadian Prime Minister Mackenzie King promptly agreed that Canada would provide the mooring mast necessary for a trans-Atlantic flight of one of the new airships to Canada, and the Canadian government set about developing the required infrastructure. In 1927, a government committee advised that a site near St. Hubert, outside Montreal, offered the most advantages. It was a 600-acre parcel of flat land that was well served by road and rail. Water and electricity were readily available, and there were no hills or trees in the vicinity to obstruct an incoming or outgoing airship. The federal government promptly acquired the St. Hubert site, ordered the mast and fittings from an English manufacturer, and waited for the first airship to arrive.

The R-100 left Cardington on the afternoon of 29 July 1930. The flight across the Atlantic went almost without incident, the ship passing Newfoundland on the evening of 30 July. Over the mouth of the Saguenay River, the R-100 hit a brief squall that was enough to cause tears (one larger than a double-decker bus) in the outer cover of the tail. The crew had no choice but to attempt repairs as the ship cruised slowly up the St. Lawrence. She offered a stirring sight to the 40,000 people who had packed Quebec City's Dufferin Terrace. Around 2:30 A.M. on 1 August, the R-100 was sighted over Montreal; for the next three hours she drifted slowly towards St. Hubert. At 5:37 A.M., after nearly seventy-nine hours in the air, the ship was secured to the mooring mast. For the next thirteen days, the R-100's visit was *the* story in Canada. The ship's crew was feted in banquets and receptions, and some 3000 dignitaries, including MPs, industry leaders, and aviation personalities, were given a tour of the airship. The hundreds of thousands of spectators who converged on St. Hubert had to be content with watching from a distance, but it was enough. On the evening of 10 August, the R-100 began a triumphal tour of Quebec and Ontario. With eighteen Canadian military and press officials as passengers and before an estimated crowd of 200,000, the airship slowly pulled away from the mooring mast at St. Hubert and sailed regally towards the setting sun. She reached Ottawa a little before 10 P.M., to the cheers of 20,000 spectators on Parliament Hill and tens of thousands more who filled nearby vantage points. Then it was on to Carleton Place, Smith's Falls, Kingston, Belleville, and Peterborough. Toronto was next, just before dawn, and for the rest of the day, she cruised majestically over eastern and southern Ontario before returning to

Montreal. And then she was gone, slipping the mast at St. Hubert on the evening of 13 August. With the exception of the torn fin, it had been a flawless flight.

The R-101 wasn't so fortunate. Early tests had not gone well, for she was desperately short on lift. The designers adopted the dangerous expedient of chopping the airship in half and inserting an extra bay, but were not given enough time to test the modifications, particularly in rough weather. Those were precisely the conditions that the R-101 encountered on its maiden flight to India, which began on the evening of 4 October 1930. Gusting winds whipped the rain into lashes as thunder rumbled through the heavy clouds and flashes of lightning lit up the countryside. As the crew battled the elements on their way to the first stop, Ismailia, in Egypt, the airship's flaws were becoming apparent. Barely thirty-five miles out, an engine broke down; nearly four hours passed before the crew fixed it and returned the ship to full power. The craft also felt very unstable. The first officer had reported soon after the test flights that the R-101 seemed poorly balanced; on this flight, the crew had difficulty keeping the nose up, and the ship frequently went into sharp dives for no apparent reason. More worrying was the fact that, despite the modifications, the R-101 did not have enough useful lift. The fat gas bags rubbed against the internal frames, and it was not long before gas was escaping through dozens of holes. With each mile, the R-101 became less buoyant—and heavier, because the outer skin was not sufficiently impermeable; instead of the rain running off, it soaked into the fabric, adding to the weight and putting a tremendous strain on the aluminum frames.

As the dirigible lurched over the Pas de Calais, its situation was desperate. Just after 1 A.M. on 5 October, it passed over Poix, in northern France, sailing into the teeth of fifty-mile-per-hour winds that drove rain into the fragile skin. Worse, the ship had been unable to regain altitude lost in the dizzying dives. By 2 A.M., the ship had reached Beauvais, where a poacher watched in fascination as it lurched towards him. The R-101 dipped lower and lower and then, at five minutes past the hour, its nose struck a small hill and exploded. Two more explosions followed, and within seconds the giant ship was a raging inferno. Six crew members clambered from the wreckage; everyone else on board, including some of Britain's most experienced airshipmen, perished. In the tragedy, the success of the R-100 was quickly forgotten. The R-101 died at Beauvais, and so too did Britain's appetite for lighter-than-air flight. The imperial airship scheme was quietly declared a dead letter, and the St. Hubert mast was never used again. It was removed in 1938 as a hazard to aircraft.

The age of the airship would limp on until the crash of the great German dirigible the *Hindenburg* at Lakehurst, New Jersey, in 1937, but eyes had already turned to the airplane as the solution to the problem of long-distance air services. Since its creation in 1927, PanAmerican World Airways, better known as PanAm, had been developing services on routes in central and South America, China, and the Pacific, and by the mid-1930s, as its Martin and Boeing flying boats gained range and reliability, it started to look longingly at trans-Atlantic passenger services. Canada could not afford to be complacent. At a conference in Ottawa in November 1935, delegates from Canada, Britain, Ireland, and Newfoundland came together to address PanAm's planned expansion into Alaska, the Arctic, and Newfoundland. The delegates agreed that Imperial Airways, Britain's national airline, would get a fifteen-year monopoly on landing rights in Newfoundland, so that Imperial could begin a trans-Atlantic service; the privilege would be extended to Canada's national airline when it came into being. Then, to avoid a battle that Imperial didn't think it could win, the delegates agreed to present PanAm with a two-stage proposal: preliminary experimental flights, to be followed by two regular flights a week for three months. If it was successful, a regularly scheduled service with stops in Shediac, New Brunswick, Botwood, Newfoundland, and Ireland would begin. PanAm and the Empire consortium should have reciprocal privileges on each other's routes; the American and British services would begin simultaneously, to ensure that neither side got a head start.

Britain's new long-range flying boats took to the air in early 1937. The first two aircraft, the *Caledonia* and the *Cambria*, were to be used on Imperial Airways's north Atlantic run; PanAm would use the Boeing 314 *Clipper*. On 5 July 1937, as PanAm's *Clipper* was leaving its base in Port Washington, New York, the *Caledonia* departed from Foynes, Ireland, bound for Botwood and Montreal. It carried no passengers or freight; every nonessential, even the floorboards, had been stripped out to save weight. The *Caledonia* and the *Cambria* would make five round-trip Atlantic flights that summer, creating a sensation in Canada similar to that caused by the arrival of the R-100. Government officials estimated that 3.5 million people saw the aircraft on their demonstration flights.

Imperial Airways mounted further test flights in 1938, but problems with new equipment delayed the program and PanAm was first out of the gate, beginning regularly-scheduled trans-Atlantic service in June 1939. Carrying a full complement of luminaries, including senators, congressmen, and officials of the Civil Aeronautics Authority, the *Yankee Clipper* left Port Washington at 1:21 in the afternoon on 24 June. On 28 June, after a three-day layover in Shediac and twenty-six hours in the air, the flying boat touched down at Southampton. The PanAm service was everything aviation boosters had hoped: speedy, reliable, safe, and, above all, opulent. For sheer comfort, the trans-Atlantic flying boats came close to rivalling the R-100. The passengers' quarters were less spacious, but white-coated stewards served gourmet meals at tables set with crisp linen, fine china, and fresh flowers. One journalist likened the rarified experience to "a yachting party in the air." But there was scarcely time to enjoy the party. In September 1939, for the second time in a generation, the world went to war.

The Second World War

On 1 September 1939, German troops rolled across the border into Poland, the dive-bombers that supported them affirming that air power would be critical in this war. Within days, the Polish armies were in full retreat and remnants of units streamed into Warsaw. The German air force, the *Luftwaffe*, had experience with aerial bombing, having mounted a controversial raid against the Spanish town of Guernica in 1937. With that in mind, it ordered a series of incendiary raids on the Polish capital. The most devastating raid came on 25 September, when the *Luftwaffe* dropped more than 600 tons of bombs on Warsaw. Two days later, its baroque palaces, cathedrals, the Royal Castle, and the old, walled town in ruins, the city capitulated. In the spring of 1940, the Nazis would turn their attentions to Western Europe, and Rotterdam would be the next major city to feel the weight of *Luftwaffe* bombs. Then, with France and the Low Countries in German hands, it was Britain's turn.

Collection of Jonathan F. Vance.

Figure 1.8 *During the Second World War, aviation was a favorite subject for children who collected bubble gum and cigarette cards.*

Beginning on 7 September 1940, London endured fifty-seven consecutive nights of bombing before the *Luftwaffe* widened the attack to industrial centres, ports, and historic cities. On 14 November 1940, ten hours of bombing left the centre of Coventry ablaze, its cathedral a ruin, its transportation and utilities infrastructure shattered. Its large aircraft factories made Coventry a legitimate military target, but public outrage at the damage to the historic city was heightened by German boasts that the *Luftwaffe* could "Coventrize" any target it wished. Attacks on English cities would continue until the spring of 1941, when the *Luftwaffe* withdrew its bomber force for the invasion of Russia. Some 40,000 civilians had been killed and a million homes destroyed, but the bombing had failed to achieve a decisive result. Britain was still in the war, and was ready to carry the air battle to Germany.

The RAF mounted many bombing raids in 1940 and 1941, but did not do a great deal of damage. Britain's bombers, the majority of them two-engine machines in service since the mid-1930s, were slow, ungainly, and carried only small bomb loads. Navigational and bomb-aiming aids were so primitive that hitting a specific target depended almost entirely on luck. Indeed, a 1941 report presented to the British War Cabinet revealed that, when attacking Germany's industrial Ruhr Valley, only one in ten aircraft was able to get within five miles of the target. There was little hope of achieving a decisive result with those odds.

But in February 1942, Bomber Command was given a new commander: Arthur Harris, a big, gregarious Rhodesian utterly convinced that the bomber could win the war. Harris had a number of advantages over his predecessor. The big four-engine bombers, the Stirling, the Halifax, and the Lancaster, were becoming available in large numbers, giving Harris the increased bomb-carrying capacity he so desired. New navigational and bomb-aiming aids would permit air crews virtual pinpoint accuracy. The United States had entered the war in December 1941, and its Eighth Air Force was building up strength after its first raid, allowing Harris to look forward to the day when Allied bombers could hit German cities around the clock.

Harris's first big raids, on Lübeck and Rostock in March and April 1942, were mere preludes to what he perceived to be the real opening of his campaign: a thousand-plane raid. On 30–31 May 1942, he scraped together every available aircraft in the British Isles, and sent 1043 two- and four-engine bombers to Köln for Operation Millennium. From Harris's point of view, the raid was a complete success; 600 acres of the city were burned and more than 18,000 buildings destroyed, rendering 60,000 people homeless (or de-housed, to use the contemporary term) and vindicating Harris's faith in bombing. The British government, having almost no other means to strike at the Nazis, gave him virtual carte blanche to bring Germany to its knees.

Over the next three years, the air offensive would demonstrate the immense destructive capacity of the airplane to a degree that made Guernica, Warsaw, and even the Blitz appear inconsequential. In July and August 1943, Hamburg was subjected to four nights of bombing that left the city centre a charred ruin and killed as many as 50,000 people—most during the raid of 27–28 July, when high explosives and incendiaries created the first man-made firestorm. Harris then sent nearly 10,000 heavy bombers to Berlin between November 1943 and March 1944, but the results were inconclusive. Not so the last big raid of the European war, on Dresden between 13 and 15 February 1945! The fabled medieval city had escaped damage thus far, but in three successive raids, 1300 aircraft of Bomber Command and the Eighth Air Force again used the deadly combination of high explosives and incendiaries to set the city alight, killing as many as 60,000 civilians. Three weeks later, American B-29 Superfortresses opened the fire-raid offensive against Japan with an attack that devastated Tokyo, killing 80,000 people and rendering a million homeless. Over the next six months, the B-29s would destroy Japan's principal industrial cities and many smaller ones in attacks far more horrifying than anything that prewar science fiction could have imagined.

The strategic bomber offensive, to which Canada contributed fifteen squadrons, represented only a part of Canada's air effort during the Second World War. Canadian fighter pilots defended Britain during the Battle of Britain in 1940, flew in support of Allied operations in North Africa, patrolled over the jungles of Southeast

Asia, and helped sweep the *Luftwaffe* from the skies over northwest Europe after the Normandy invasion. Long-range bombers and flying boats of the RCAF guarded convoys in the North Atlantic and stood watch on the approaches to the west coast. Canadian airmen could be found on air bases around the world, from the Cocos Islands in the Indian Ocean to Reykjavik in Iceland, from Takoradi in western Africa to Murmansk in northern Russia.

Military aviation was not the only beneficiary of the war; virtually every other aspect of aviation had expanded as well. By September 1939, Trans-Canada Airlines had only the beginnings of a national service. It flew over three million miles in 1939, but carried fewer than 22,000 paying passengers. After six years of war, the airline was transformed. Its workforce exploded, from fewer than 500 in 1939 to more than 3200 in 1945; in the last year of the war it carried more than 180,000 passengers and flew some 11.5 million miles. The fleet, which began with two Lockheed Electras, sleek, ten-passenger aircraft that were the last word in airliners in 1937, grew to twenty-eight aircraft, including bigger, more modern machines such as the Lockheed Lodestar and the DC-3. New routes were added—Moncton in 1940, Halifax in 1941, Sydney and St. John's in 1942, and Victoria in 1943.

But TCA was not the only player. The Canadian Pacific Railway, still smarting from being squeezed out of the national airline business in 1937, had retained an interest in aviation through its partial ownership of Canadian Airways Limited, and the railway recognized that wartime conditions created an opportunity to increase its involvement. Beginning in 1940, the CPR aggressively acquired aviation companies operating in the Canadian north and, in May 1942, amalgamated its acquisitions, as well as CAL and its affiliates, into Canadian Pacific Air Lines. A year later, the new carrier was flying 18,000 miles a day, deriving fully 60 percent of its revenue from passenger fares. CPAL had one major intercity route, a daily service between Montreal and Quebec City; the bulk of its flying was in the northwest, from its hubs at Edmonton and Winnipeg. The airline's advertisements made much of the fact that it complemented rather than competed with TCA and that together the two airlines gave Canada a passenger and freight system second to none.

Collection of Dallas Hart.

Figure 1.9 *Luxury flying, 1940s style—inside the cabin of a TCA Lockheed.*

What was it like to fly in those days? In short, it was unpleasant, beginning with the reservation process. Everything was done in Morse code between TCA stations, and it could take up to twenty-four hours to get a seat confirmed; someone flying on short notice often had to leave on the first leg of a journey without knowing if there were seats available on subsequent legs. Then you had to pay, and it was not cheap—a one-way ticket between Ottawa and Vancouver cost $145 in 1939, or over $2100 in current values. Weather-related delays and cancellations were frequent, and TCA's Lockheeds were buffeted by so much turbulence that airsickness was common. The cabins were not pressurized and flights generally occurred at between 5000 and 8000 feet, so passengers frequently had to make use of the cabin oxygen masks. Propeller noise meant that it was often impossible to carry on a conversation, and toilet facilities were primitive—a crank opened a hole in the bottom of the toilet, and waste was sucked out by the slipstream.

Angels of the Air

The saving grace of passenger flying in those days were the Angels of the Air. United Airlines had hired the first stewardesses in 1930, and TCA followed suit in 1938 when it put Lucille Garner in charge of setting up its stewardess department. The requirements were stringent: under 5'4" in height (because cabin ceilings were so low) and strong enough to carry baggage and wrestle the cabin door shut, girdles and stockings were mandatory, only certain colours of lipstick and nail polish were allowed, and hair had to be put up above the collar—all for $140 a month, out of which the women had to pay for their own uniforms and hotel bills whenever a flight delay or cancellation forced a layover.

Trans-Atlantic flying, a minor miracle before 1939, also became commonplace during the war. PanAm flying boats continued to cross between Europe and North America, but passenger space was in high demand and the Canadian government usually had to wait its turn behind others. Frustrated by delays, the government acquired a Lancaster bomber, which was converted to a passenger-carrier, and inaugurated the Canadian Government Trans-Atlantic Service on 22 July 1943; within a year, it and two other Lancastrians (as the modified aircraft were renamed) were making three return trips to England each week. In its first eighteen months, the service carried more than one million pounds of mail and 2000 passengers, considerably more than had flown the Atlantic between the R-34's flight in 1919 and the opening of PanAm's scheduled service in 1939.

Collection of Dallas Hart.

Figure 1.10 *An Angel of the Air—welcoming passengers aboard a TCA flight in Winnipeg, 1942.*

But the Lancastrians were not the only aircraft regularly crossing the Atlantic. Before the war, the British Air Ministry had used North American manufacturers to build up its squadron strength, and British orders increased dramatically through 1939 and 1940. But it was not clear how those aircraft (26,000 on order by the summer of 1940) would get to Britain. Taking them apart for transport by sea was time-consuming and wasteful—after all, they had just been put together. Why not fly them over? The British Air Ministry thought the plan was suicidal, given the recent history of trans-Atlantic flying, but it appealed to Lord Beaverbrook, the Canadian-born Minister of Aircraft Production. Ignoring the Air Ministry, he struck a deal between the Canadian Pacific Railway (which would provide ground logistical support) and his own ministry, which would supply the flying personnel and pick up the tab. In November 1940, three months after Beaverbrook concluded the agreement, seven Lockheed Hudson medium bombers (the military version of the Lockheed Electra) made the first successful ferry flight.

Although the first few months were marred by a number of tragedies—the most serious a February 1941 crash that killed Nobel Prize-winning scientist Sir Frederick Banting—the proportion of completed flights was so encouraging that the program was expanded. The number of aircraft ferried across the Atlantic increased dramatically—more than 1900 in 1942, nearly twice that two years later. The stream of aircraft, flying either singly or in small groups, left the main North American base of Ferry Command in Montreal, bound for Prestwick in Scotland, via either Gander or Goose Bay and Reykjavik. By 1944, an average of ten aircraft left Ferry Command bases each day to cross the Atlantic. The flight was not always a pleasant one, but the loss rate was remarkably low—just 2 percent, fewer than 200 of the nearly 10,000 aircraft flown across the Atlantic by Ferry Command. (Only about a third of the lost aircraft went down in mid-ocean; the majority crashed over land, many of them on takeoff or landing.) The success rate of Ferry Command demonstrated the exceptional skill of the crews who routinely retraced the route of Alcock and Brown, and Lindbergh. Without their deliveries, the Allied air forces would soon have been starved of equipment.

Canada's other great contributions to the Allied war effort was the British Commonwealth Air Training Plan. Training aviators in Canada had been proposed as early as 1936, but intermittent negotiations failed to produce an agreement. The coming of war, however, had a salutary effect on both sides. The British made a new formal proposal on 6 September 1939, before Canada entered the war, and Mackenzie King responded with an enthusiastic pledge to expand training facilities for airmen. On 17 December 1939, Britain and Canada signed the agreement for the British Commonwealth Air Training Plan. Intended to produce 1464 pilots, observers, and wireless operators / air gunners each month, the scheme would require an immense infrastructure. Initial estimates put the requirements at 3500 aircraft (more than ten times as many as the RCAF had when the war began), 33,000 servicemen and women, 6000 civilian employees, and more than a hundred airfields. After an exhaustive survey of Canada's 153 registered airfields, officials calculated that twenty-four of them were ready or could be made so in fairly short order; another dozen required extensive upgrading; and the remaining seventy-five airfields would have to be built from scratch. By war's end, what American President Franklin Delano Roosevelt called "the aerodrome of democracy" had produced nearly 132,000 graduates from more than one hundred training establishments employing tens of thousands of military and civilian personnel.

Wartime Aircraft Manufacturing

Between 1919 and 1939, a total of 678 aircraft of all types were manufactured in Canada by a small number of specialized firms—Fleet Aircraft of Canada, a tiny operation in Fort Erie, Ontario, with a full-time staff that could fit in one of TCA's Lockheeds, and the larger Noorduyn Aircraft of Montreal, which produced the legendary Norseman bushplane—and a few big concerns, such as Canadian Car and Foundry, National Steel Car in Hamilton, and the Ottawa Car Manufacturing Company, which manufactured British-designed aircraft as a sideline. When demands for military aircraft skyrocketed, Canadian manufacturers were quick to increase production. In Fort William, Ontario, and Verdun, Quebec, CCF manufactured Hurricane fighters and the Hampden, a medium bomber adored by its crews despite being obsolete, on an assembly line that was the brainchild of Vancouver-born Elsie MacGill, the first woman to graduate in electrical engineering from the University of Toronto and the first woman to take a master's degree in aeronautical engineering from the University of Michigan. Victory Aircraft in Malton, Ontario, was running two nine-hour shifts

Figure 1.11 *An Australian-crewed Handley-Page Hampden over the North Sea, 1942. Aircraft like this one were manufactured in fort William, Ontario (now Thunder Bay), and Verdun, Quebec, by Canadian Car and Foundry.*

six days a week to turn out fifty Lancasters a month. De Havilland in Downsview produced the Tiger Moth, a graceful and much-loved training biplane, and the Mosquito, a revolutionary high-performance medium bomber constructed primarily of wood. Canadian Vickers at Cartierville, Quebec, and Boeing's Canadian plant in Vancouver built Canso flying boats, used for antisubmarine patrols over Canada's coasts. By the end of the war, Canadian workers were building more aircraft in a month than the prewar aviation industry had built in a decade.

The growth of the BCATP was part of a remarkable expansion of military aviation during the Second World War. The manufacturing sector exploded to employ some 116,000 workers by 1944, and by war's end they had built over 16,000 fighters, bombers, long-range flying-boats—the most sophisticated military technology the world had ever seen. From a minuscule service of 270 aircraft and 3048 all ranks in 1939, the RCAF emerged in 1945 with 249,662 men and women having worn the air force blue, and at its peak strength in January 1944, the RCAF boasted seventy-eight squadrons, thirty-five of them overseas. Canada was a true aviation superpower, with the fourth largest air force in the world.

A Return to Peace

When the Second World War ended in 1945, the exponential growth of all sectors of the aviation industry over six short years meant that flying was no longer a mystery to be practiced by the very few. It had become the very thing that the air lobby had been looking forward to for thirty years: a normal, unremarkable part of life. But in that transformation, something was lost—the magic that had fascinated aviation pioneers was becoming a thing of the past as flying became bigger, more commercial, more reliant on technology, and more critical to international economics.

It began in the cockpit, where war had brought new devices to enhance the efficiency and effectiveness of aircraft. The Air Position Indicator and the Ground Position Indicator, both primitive versions of GPS, allowed a navigator to plot, with reasonable accuracy depending on wind conditions, the aircraft's position. Oboe, Gee, H2S, and

G-H were navigational aids that relied on either ground-based radio beams or airborne radar sets. An automatic pilot system, known as George, maintained an aircraft in straight and level flight virtually indefinitely. These devices made aircraft safer and more efficient, but they also reduced the role of the pilot in the operation of the airplane. This led to the apocryphal pilot's lament that would have been unthinkable a generation earlier: the pilot on a scheduled carrier was simply a bus driver, on a charter service simply a taxi driver.

But flight assistance devices were only part of the story. Through the war, airplanes had become larger, and correspondingly more complicated to build, maintain, and operate, particularly with the advent of the jet engine. Early aircraft pioneer Glenn Curtiss had started as a motorcycle manufacturer, because the differences between a motorcycle engine and an airplane engine were minimal. Even with the most sophisticated engines of the Second World War, the difference between a 1918 engine and a twelve-cylinder Rolls Royce Merlin engine that powered countless Allied warplanes was only one of scale. But with jet aircraft coming on stream, everything changed. Mechanics needed a new set of skills, and a new set of tools—and that was expensive. Success in the new aviation world would depend on a strong economy.

After the First World War, Canada's economy had nose-dived as the accelerated spending of the previous four years was abruptly curtailed. In 1945, Canadians wondered if the same thing would happen after Germany and Japan were beaten. But after a few minor hiccups, the economy continued to grow. Pent-up demand fuelled the resurgence of consumer manufacturing, and industry kept employment levels high as a result. Bigger, more interventionist government meant that there was plenty of money to go to new programs. Although the RCAF was slashed by some 90 percent, to a regular force of just over 16,000 personnel and eight squadrons (although those numbers would increase again as Cold War tensions deepened) and the BCATP bases were closed, there was considerable growth in many other areas, For the aviation sector, those were heady days.

Aerial photography continued, with the RCAF committing to an ambitious five-year plan to photograph every corner of Canada. In 1946, Saskatchewan inaugurated an air ambulance service when a war-surplus Norseman flown by an ex-RCAF instructor ferried a woman suffering from acute diabetes from a remote farm to a Regina hospital. For nearly a decade, this was the only organized, publicly funded air ambulance service in the western hemisphere. And the aviation manufacturing sector retooled to supply civilian demand. In July 1946, Canadian Vickers in Montreal, which had become Canadair in 1944, introduced the North Star, Canada's first modern passenger airliner, albeit one converted from a war surplus American DC4. The post-war generation got used to travelling by air, so traffic grew, and grew dramatically. In the fifteen years after the end of the Second World War, traffic increased by a factor of ten, with the U.S. aviation sector leading the way. By 1961, Aeroflot was the largest airline in the world, but the next five were U.S. carriers; altogether U.S. airlines carried 52 percent of the world's air traffic. Would this lead to ruinous competition, or would there be freedom of the skies?

The first generation to fly had recognized the need for regulation, and eighteen European governments came together in Paris in 1910 to lay the groundwork for an international aviation convention. But with so little flying going on, the need for regulation did not seem pressing—until the First World War. An international aviation committee was established in 1918, and in 1919 thirty-eight governments ratified the International Air Convention, an agreement of forty-three articles dealing with various aspects of civil aviation. Perhaps most significant, it created the International Commission for Air Navigation, an organization that remained in existence until 1944. In that year, after the U.S. government convened an international conference on civil aviation in Chicago, it was replaced by the International Civil Aviation Organization. (The first woman to serve as technical advisor with the ICAO was none other than Elsie MacGill, who went on to hold senior engineering posts with some of the biggest aircraft manufacturers in Canada.) Headquartered in Montreal (an indication of the critical importance of Canada in contemporary aviation), the ICAO was envisioned as a means to ensure consistency in regulations, standards, and procedures of civil aviation—everything from the licensing of ground and air crews and aircraft to communications and air traffic control. Although there were still many points of contention between the world's aviation superpowers, for the moment Ottawa had good reason to imagine that Canada's prominent position in the air was secure. But the fate of the legendary Avro Arrow revealed the hazards of the high-cost world of modern aviation.

Collection of Jonathan F. Vance.

Figure 1.12 *Stamps issued to commemorate Canada's heritage in military aircraft manufacturing (clockwise from top left): the Avro Canada CF-100, the Avro Lancaster, the Curtiss JN-4 Canuck, and the Hawker Hurricane.*

The great British aircraft manufacturer A.V. Roe had been casting around for business opportunities after the war, and decided to purchase Victory Aircraft Limited, the Malton, Ontario, factory that had built Lancaster bombers during the Second World War. The new entity, A.V. Roe Canada, or simply Avro, was determined to continue its tradition of innovation, and debuted in 1950 the Avro Jetliner, an aircraft that was a technological success but a commercial failure. Despite its clear superiority over anything else on the market, the Jetliner could find no buyers, in Canada or elsewhere, and the prototype was later sold as a plaything to American billionaire Howard Hughes.

Collection of Eddie Coates.

Figure 1.13 *The elegant and superbly designed Avro Jetliner.*

Then Avro turned its attention back to military aviation, with a design for the CF105, intended to replace the RCAF's fleet of aging CF100s. Designated the Arrow, the twin-engine, all-weather, two-seat fighter went into development at Malton, with the prototype rolling off the line in October 1957. Its first flight, on 25 March 1958, exceeded all expectations. Its speed, rate of climb, range, and all-around performance put it years ahead of anything else on the market, and that was before the specially designed engine package was mounted. Avro waited for buyers to beat a path to its door, but instead the government of John Diefenbaker dropped the axe: on 20 February 1959, it announced that the Arrow program would be scrapped. For many people, at the time and since, it was either a blunder of epic proportions that robbed Canada of the chance to lead the aerospace industry, or a near-criminal boondoggle, with the U.S. defence industry forcing Ottawa to kill the program because, overnight, the Arrow made every U.S. competitor obsolete.

The real story is less controversial. In the first place, the Russian launch of its orbital satellite Sputnik in 1958 had changed the defence game. Military observers decided that future attacks on North America would be launched, not by conventional bombers, but by intercontinental ballistic missiles against which a fighter aircraft, even one as good as the Arrow, could offer no defence. But a more compelling argument was economic. At the beginning of the Arrow program, estimates put the cost at $1.5 to $2 million per aircraft, based on initial sales of 600 units. But Avro's cost overruns were staggering. By 1959, the costs had risen to somewhere between $8 and $15 million (depending on how costs were calculated), and there seemed to be no reason why the costs would not climb even higher once the Arrow went into production. Even at a conservative estimate of $15 million per aircraft, the RCAF's 100 Arrows would cost $1.5 billion—at a time when an off-the-shelf F86 Sabre (admittedly a much inferior aircraft), manufactured at the Canadair plant in Montreal, could be bought for $250,000. To put it in another context, until 1970 defence was the biggest single item in the federal budget. Between 1949 and 1959, the federal government sank $960 million (over $7 billion in current values) into the Arrow program—more than it had spent on external affairs, education, welfare, and health. If the Arrow program were to continue, it would mean that every other part of the federal government would have to be cut back, in some cases to nothing. In reality, the Diefenbaker government had no alternative to drop the boom—any other course would have been irresponsible, even when a mythical aircraft like the Arrow was at stake. Canada would go on to have other triumphs in aircraft manufacturing—particularly in the regional and corporate jet sectors with Canadair's (later Bombardier's) Challenger series, De Havilland's (later Bombardier's) Dash series, and Bombardier's Regional Jet series—but military aviation increasingly came to be dominated by multinational consortia, such as Panavia and EADS. It was simply too expensive a game for a middle power like Canada.

The Future of Canada's Aviation Industry

During the 1940s, as Canada's system of national air routes developed, the two strongest players were clearly TCA and Canadian Pacific Air Lines. But while TCA had its business handed to them by the federal government, CPAL had fought for every route. It clawed its way into the airline business, first working small-town and remote routes, largely in western Canada, and in 1947 was rewarded with government approval to operate a significant intercity route, the run between Vancouver and Calgary. This was enough for CPAL's new president, Grant McConachie, to build on. With most of TCA's attention focused on routes to Europe, McConachie demanded that the government give him the rights to the Pacific. Ottawa agreed, but at a price: CPAL would have to operate a service between Canada and Australia before it would be allowed to serve other parts of the Pacific Rim. It was an underhanded move; the government believed there was no way the Australian route could be made to pay, and CPAL would probably go bankrupt trying. And it almost did. The flight from Vancouver to Sydney was thirty-seven hours and required many stops along the way; each flight cost the airline $18,500, so to break even it had to carry twenty-seven passengers. It almost never did. Partly because of the cost, which was $685 one-way to Australia (nearly $7000 in current values), it was not unusual for the flight to leave Vancouver with only a single passenger on it. But Grant McConachie was saved by a couple of lucky circumstances. With the Communist revolution in China, Hong Kong was thronged with wealthy refugees looking for a way out; they found it on CPAL's North Stars, and the airline could fill as many flights as it could lay on. Then, when the Korean War began in 1950, there were lucrative government contracts to fly military personnel back and forth from North America to Southeast Asia. Revolution and war proved to be the salvation of CPAL.

Figure 1.14 *A Canadair North Star of Canadian Pacific Airlines. In 1954 it crashed after a mid-air collision with a training aircraft, killing all on board.*

There were more bumps along the way, some related to aircraft choice (CPAL acquired two of the worst British airliners of the postwar period, the Comet and the Britannia), but the carrier's future was apparently secured with the Transportation Act of 1967, which institutionalized a completely new system of regulation in the aviation sector. Now, there would be two national carriers: Air Canada (as TCA had been renamed in 1965) and CPAL, now more commonly known as CP Air. There would be no competition over internal routes; instead, each airline would be granted a monopoly over certain flights, with CP Air receiving 25 percent of the routes and Air Canada the rest. International routes were also divided up, with Air Canada getting the rights for Britain, central Europe, and the Caribbean, and CP Air claiming South America, the Far East, Australia, and a single European destination, Amsterdam. Rights to U.S. destinations were parcelled out on a piecemeal basis.

The Transportation Act represented a major accomplishment on the road to profitability for Canadian carriers, but it occurred within a framework of escalating costs. Part of those costs came with the advent of the Boeing 747 Jumbo Jet, which first reached Air Canada in 1971. Canadian variants were configured to seat 365 passengers, which meant that the 747 could be enormously profitable—on routes that had high occupancy rates. But aggressive expansion meant that high occupancy rates were hard to find. Air Canada bought up local airlines to act as feeders for its main interurban services, a good idea in theory but one that had significant consequences in terms of debt. Furthermore, Air Canada was privatized in 1988, and faced a future outside of the shelter offered by government ownership. CP Air ran into greater problems in this regard. In 1986, it was bought by Pacific Western Airlines, which integrated the carriers into Canadian Airlines International. Then, in 1989, it bought Wardair, Canada's third largest carrier; once a bush airline flying out of Yellowknife and Edmonton, it eventually got into the charter market to become the major Canadian carrier serving holiday destinations like Hawaii and Barbados. Wardair was a good airline with strong passenger loyalty and an excellent safety record, but it had something that the new Canadian Airlines International coveted: the rights to fly to the UK. From CAI's perspective, it was a good buy, but it increased the airline's already crippling debt load, and gave them lots of aircraft they couldn't use at a time when they were already struggling with overcapacity.

Figure 1.15 *One of Wardair's promotional postcards, from the happier days when the airline was one of the success stories of Canadian aviation.*

The fact that Air Canada and Canadian were both facing heavy debt at the same time would not have been so significant had it not coincided with the period when both airlines had to upgrade their fleets. Regular maintenance was costly enough. In the late 1970s the seats on a 747 cost about $1000 each, and Wardair's 747s were fitted out to carry 456 passengers. With an eighteen-month replacement schedule, each Jumbo needed half a million dollars worth of seats every year and a half. Fire-retardant, noise-absorbing, extra durable wallpaper cost about $300 a panel, which translated into tens of thousands of dollars each time an airplane's wallpaper needed replacing. All of this meant that passenger airlines were a very low-margin business, depending on the route. In the late 1970s, some of Wardair's northern routes were paying about 31 cents per passenger per mile, a very decent return, but the carrier's charter operations struggled to pay 8.1 cents per passenger per mile. When the issue was fleet replacement rather than scheduled maintenance, costs skyrocketed. In 1988, Air Canada ordered thirty-four new Airbuses at a cost of $1.8 billion, making it the biggest aircraft purchase in Canadian aviation history. A few years later, it laid down another $2 billion for twenty-four Canadair Regional Jets and five more Airbuses.

That fleet expansion took place within the context of deregulation. The federal government had long been a powerful player in the airline industry in Canada, but by the 1970s this role was coming under increasing scrutiny. Ottawa had built an enormous infrastructure for the supervision of the airline industry, despite the fact that in 1984, only 4.6 percent of all intercity passenger trips were carried out by air. In simple mathematical terms, this meant that the majority who didn't fly subsidized the minority who did. The prime impetus for deregulation, then, was cost-cutting. In April 1987, for example, Ottawa announced a new policy for airport management in Canada in which local authorities could take over ownership and operation of federal airports. In 1984, Canada's minister of transport announced that the federal regulatory body should "give much greater weight to the benefits of increased competition in judging the requirements of public convenience and necessity." Next came the new National Transportation Act, which stipulated that the granting of a licence would be virtually automatic, as long as the applicant was at least 75 percent Canadian owned or controlled.

It is difficult to assess the immediate impact of deregulation, because in the early 1990s, a number of crises hit Canada's aviation sector. The Gulf War of 1991–92 drove up fuel prices even as greater competition between carriers was driving down income. Passenger numbers were up, but so too was the number of flights, which had a ruinous impact on the bottom line. Flights that in the early 1980s had averaged 70 percent of capacity now dropped to under 65 percent of capacity. Losses at CAIL mounted, rising from $56 million in 1989 to $543 million in 1992; that year, the airline had to lay off 2000 workers. In November 1992 the federal government announced that it wasn't interested in bailouts to help either of Canada's major carriers, insisting that the market should shape the future of air travel in Canada. This was more symbolic than anything, because both levels of government still provided loan guarantees to keep the airlines afloat, but the symbolism did nothing to help CAIL. Its share price, which had been about $250 in 1989, dropped to just over a dollar in 1996. Eventually, it could carry on no longer and was bought by Air Canada in 1999. The main consequence of the merger for passengers was a reduction in service. Domestic flights by the new Air Canada group were reduced by nearly 25 percent between August 1999 and August 2000; new flights were eventually added by regional carriers, but the net loss after a decade was still about 12 percent of domestic flights. There were also fewer aircraft in service (from 423 before the merger to 375 after), which perhaps wasn't a bad thing given the overcapacity in the Canadian industry.

Two years later, another thunderbolt hit the aviation world with the terrorist attacks of 9/11. Whatever else it did to the world, the image of passenger aircraft slamming into the towers of the World Trade Centre in New York dissuaded millions of people from travelling by air. In Canada, passenger numbers went from 80 million in 2000 to 71 million in 2002, a drop that the already stretched carriers couldn't afford. The response was cutbacks—some 10,000 jobs lost in the Canadian aviation sector after 9/11—and a huge increase in the number of airlines moving towards the precipice of ruin. Of the six major U.S. carriers at the time (American, United, Delta, US Airways, Continental, and Northwest), Continental seemed to be in the best shape: its break-even point after 9/11 was 77 percent, meaning that its airplanes had to be 77 percent full to make a profit. The break-even point for United was a staggering 96 percent, so the airline was losing money on virtually every flight.

In the first decade of the twenty-first century, the fortunes of Canada's aviation sector are mixed. The darks clouds hanging over Air Canada as it battles towards profitability must be balanced against the success of Calgary-based Westjet, with its no-frills flights and an emphasis on friendly service. Bombardier has had to make massive layoffs in the economic crunch of 2009, but few analysts doubt that this is anything but temporary. With its products still held in high regard around the world, it will in all likelihood rebound with the economy.

The motto of the Royal Canadian Air Force was *Per Ardua ad Astra*—through adversity to the stars. It could just as easily serve as the motto for Canadian aviation in general. There has been much adversity in this history, and anyone who argues that modern-day challenges are unprecedented should look more carefully at the past. In the early 1920s, the vast majority of Canada's tiny flying companies lived on the edge of collapse, working from contract to contract. It was no different in the 1930s, when small aviation firms came and went with depressing regularity. Still, we must admit that, ever since Doug McCurdy muscled the *Silver Dart* off a frozen lake in Nova Scotia in 1909, Canadians have always found a way to fly. One suspects they always will.

Name _____ **Date** _____

PRACTICE ACTIVITY

Throughout this chapter, the contributions of several Canadian aviation pioneers have been discussed. Choose the top five people you believe have had the greatest influence on aviation in Canada. List them by rank (1–5), describe their contribution, and explain why you believe they had a significant influence on Canadian aviation.

Rank	Name	Contribution	Significance

1. _____

_____

_____

_____

2. _____

_____

_____

_____

3. _____

_____

_____

_____

4. _____

_____

_____

_____

5. _____

_____

_____

_____

CHAPTER **2**

CIVIL AVIATION

OBJECTIVES

After completing this chapter, you will:

- Understand the structure and operations of the general aviation sector in Canada
- Become familiar with pilot license progression
- Learn the importance of personal aviation and the potential impact of very-light jets (VLJs)
- Understand corporate aviation and fractional ownership
- Become familiar with helicopter operations in Canada
- Understand the sectors, management, and employee groups of Canadian airlines

Introduction

Civil aviation refers to the entire aviation industry, except for military aviation. To the general public, commercial scheduled air travel (commonly called *airlines*) is the most recognizable segment of the aviation industry. However, there are more aircraft and pilots in the general aviation sector than in the airlines. Therefore, it is important to develop a broad understanding of the various operations within the civil aviation industry in Canada.

General Aviation

Canada operates the second largest civil aircraft fleet worldwide with 29,064 fixed-wing and 2,505 rotary-wing aircraft across the country (Transport Canada 2009). The majority of these aircraft, roughly 80 percent, operate within the general aviation (GA) sector. GA refers to all air operations that take place outside of the military or airline environment.

Canadian Owners and Pilots Association (COPA)

The Canadian Owners and Pilots Association (COPA) was formed in 1952 to represent the interests of personal aviation, which is the sector of GA in which people fly for personal travel and recreation. COPA is a nonprofit organization that strives to support aviators by defending their right to the freedom of Canadian airspace. Since the 1950s, COPA has worked to raise awareness of issues that impact the flying community, promote air safety through education and lower the cost of flying.

In 1952 there were less than 7,000 pilots in Canada (4,560 private pilots and 2,025 commercial pilots) and commercial aircraft outnumbered private aircraft (1,294 to 966). This meant that the majority of private pilots rented certified-aircraft, rather than owning their own aircraft. Since that time, the number of aircraft in Canada has increased dramatically (refer to Table 2.1).

However, the majority of this increase was the result of private aircraft ownership (refer to Table 2.2). In recent years, personal aviation has grown in popularity.

As of 2008, there were 64,542 licensed pilots in Canada. Of these pilots, 39,944 were non-commercial (Transport Canada 2009). Therefore, in Canada non-commercial pilots outnumber commercial pilots 2:1. This ratio has held relatively steady over the last twenty years (refer to Table 2.3).

General Aviation (GA) Sectors

The range of activities within GA varies tremendously. However, some of the major components of GA include flight training, personal aviation (which is travel for personal reasons and recreation), corporate travel, and helicopter operations.

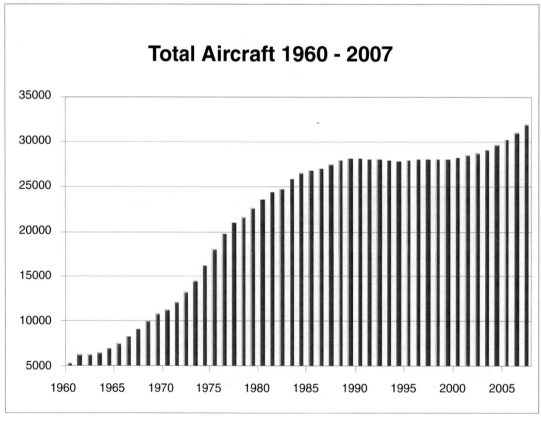

COPA 2008.

Table 2.1 *Total Aircraft 1960–2007*

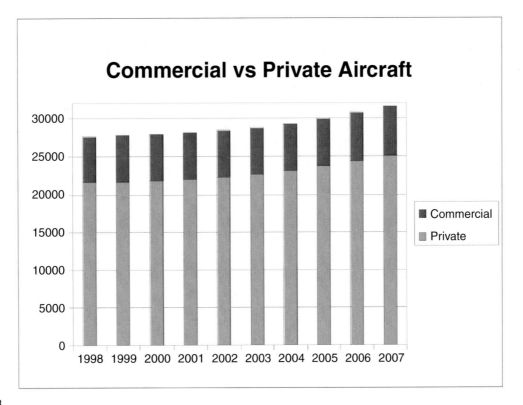

COPA 2008.

Table 2.2 *Commercial vs. Private Aircraft*

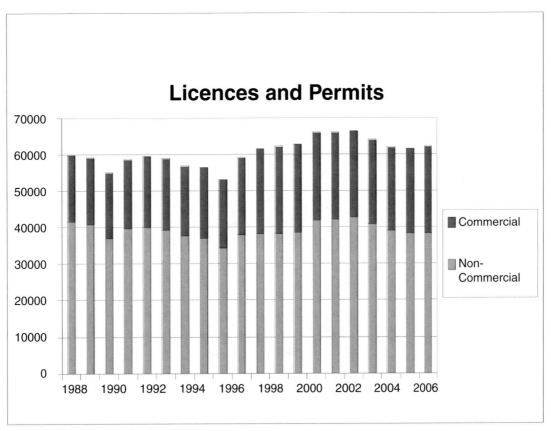

COPA 2008.

Table 2.3 *Commercial vs. Noncommercial Licenses and Permits*

Flight Training

Flight training is an important component of GA. Hundreds of local airports, which are not serviced by scheduled airline flights, are home to active flight schools and flying clubs. Across Canada, there are 388 active flight training units (Transport Canada 2009).

The path to a career as a pilot requires a tremendous amount of knowledge and skill-training combined with hours of experience. The number of flight hours a pilot has completed is directly linked to his or her career progression. For example, a pilot must have completed approximately 2000 hours to be interviewed for a position with Air Canada. Therefore, from a student pilot's first training flight, every hour is carefully documented in a pilot logbook.

There are several licenses that a pilot can attain. Pilots who want to fly for recreational purposes typically achieve a Pilot Permit–Recreational (PP-R). With a PP-R, a pilot can fly only during daylight hours and in fair weather conditions with a maximum of one passenger. However, it is a relatively inexpensive way to become a pilot. On the other hand, pilots who begin flight training with the goal of becoming a professional pilot begin their training by studying for a private pilot's license (PPL). A PPL allows pilots to fly for recreation while building their flying time towards a higher level license. This level of license does not allow pilots to fly for hire. Therefore, career pilots typically begin training for their commercial pilot's license (CPL) shortly after completion of a PPL. A CPL allows pilots to fly for hire. However, the job opportunities for new commercial pilots are very limited. Typically, a new commercial pilot has two options: (1) become a flight instructor, or (2) become a bush pilot.

Whatever option is chosen by a new commercial pilot, some form of additional training will be required. To become a flight instructor, a commercial pilot must add a flight instructor rating to his or her license. There are four classes of flight instructors, Class 4, 3, 2, and 1 (with 1 being the highest classification). A flight instructor will start out as a Class 4 and must be supervised by a Class 2 or Class 1 instructor. As a flight instructor builds experience, he or she will progress through the classifications, eventually becoming a Class 1 instructor.

On the other hand, a bush pilot typically requires a float rating, which takes five hours of training to complete. However, bush pilots often get into a frustrating situation: they won't be hired without fifty hours of float experience—yet they can't gain the experience without being hired.

Shutterstock © Karen Gentry.

Figure 2.1 *Float plane operations are common throughout Canada, particularily in the North.*

After approximately two to five years experience as a flight instructor or bush pilot, enough experience will have been accumulated to achieve an airline transport pilot's license (ATPL). An ATPL is the highest level of license a pilot can achieve and allows him or her to work as an airline pilot.

A pilot's training is never complete, as a type rating course must be completed any time an airline pilot switches aircraft, and annual recurrent training must be completed every year. Overall, flight training is an important component of every pilot's career and of the aviation industry as a whole.

DID YOU KNOW?

Only 4 percent of ATPL license holders are female.

PROFESSIONAL PROFILE

Job Title:

Flight Instructor or Bush Pilot

Qualifications and Job Characteristics

Flight Instructor: A flight instructor is a pilot who teaches others to fly. This instruction involves the teaching of ground-school in a classroom and in-aircraft flight lessons. To become a flight instructor a pilot must have his or her commercial license endorsed with a flight instructor rating.

Bush pilot: A bush pilot lives and works primarily in remote northern areas of Canada. The bush pilot may deliver medical equipment or supplies, assist with logging operations, spot forest fires, or transport personnel and equipment to areas inaccessible via roads. The advantage of bush flying is that only a commercial license is required. Therefore, new commercial pilots save the time and money that would be invested in their flight instructor training.

Salary: The salary range for entry-level bush pilots and flight instructors is $20,000–35,000/year.

Lifestyle: It is common for aviation companies to hire new commercial pilots as dispatchers or ramp workers for a year or more before they transition to instructing or bush flying. It is helpful to regard this phase in a pilot's career as a long interview. During this period of time, the company will assess the pilot's professionalism, competence, and reliability. Once a pilot begins working as an instructor or bush pilot, he or she will work very hard for very little pay. This phase in a pilot's career can be regarded as an internship, as the education gained during this period lays the foundation for the rest of his or her career. Most pilots spend two to five years in this role before they have gained enough experience to complete their airline transport pilot license and move on to fly larger aircraft.

Pilot Licenses—Quick Reference

	RECREATIONAL	PRIVATE	COMMERCIAL	AIRLINE TRANSPORT
Privileges	Fly for fun	Fly for fun or transportation	Fly professionally	Fly professionally as an airline pilot
	Flight permitted during daylight in visual meteorological conditions. May carry only 1 passenger. Only a seaplane endorsement can be added to this license.	Pilot cannot work for hire. Endorsements can be added to license to allow pilots to fly seaplanes, at night, multi-engine aircraft, etc.	Private pilot's license is a prerequisite. Endorsements can be added to license to allow pilots to fly seaplanes, at night, multi-engine aircraft, etc.	Private and commercial pilot licenses are prerequisites. Requires a group 1 instrument rating.
Minimum Age	16	17	18	21
Medical Certificate	Category 4	Category 3	Category 1	Category 1
Knowledge	60% on recreational pilot written examination covering air law, navigation, meteorology, and aeronautics (general knowledge)	40 hours private pilot ground school instruction 60% on private pilot written examination covering air law, navigation, meteorology, and aeronautics (general knowledge)	80 hours commercial pilot ground school instruction 60% on commercial pilot written examination covering air law, navigation, meteorology, and aeronautics (general knowledge)	A minimum of 70% in each of 3 written examinations covering the following subjects: 1. Meteorology, radio aids to navigation and flight planning 2. Air law, airplane operation, and navigation 3. Instrument rating

	RECREATIONAL	PRIVATE	COMMERCIAL	AIRLINE TRANSPORT
Experience	25 hours recreational pilot flight training including: • 15 hours dual instruction • 5 hours solo flight time	45 hours flight training, 5 of which may be conducted in a simulator. Must include: • 17 hours dual instruction time with 3 hours cross-country and 5 hours instrument time • 12 hours solo flight time with 5 hours cross-country	200 hours flight time with: • 100 hours pilot-in-command, including 20 hours cross-country • 65 hours commercial pilot flight training after completion of private pilot's license	1500 hours flight time with: • 250 hours pilot-in-command in airplanes, including at least 100 hours cross-country flight time with a minimum of 25 hours at night • 100 hours of night flight time • 100 additional hours cross-country flight time as pilot-in-command • 75 hours instrument flight time
Skill	Completion of a recreational pilot flight test	Completion of a private pilot flight test	Completion of a commercial pilot flight test	Completion of an airline transport pilot license flight test in a multi-engine aircraft

*Please note that this information is only an overview. Refer to the Canadian Aviation Regulations (CARs) for complete details on pilot licensing requirements. These hours represent Transport Canada minimums. In many cases, pilots will require additional flight experience before achieving license standards.

MULTI-CREW PILOT'S LICENSE (MPL)

ICAO has come to realize that the current method of training pilots is based on the way pilots have always been trained. Pilots are often taught what they need to know in order to pass a flight test, rather than what they need to know to do their job! This method does not reflect the modern understanding of how people learn or prepare them to use the advanced technology in the cockpit. This realization led to a new licensing structure that TC has endorsed, termed a multi-crew pilot's license (MPL). An MPL allows a person to be a co-pilot (not the pilot-in-command) on multi-engine transport category aircraft in VFR or IFR flight conditions. A MPL is different than traditional licenses because, upon completion of a fifteen-month training program, the pilot will be flying commercially in a passenger jet. This is unheard of within the traditional pilot licensing system. This license will require the advanced use of simulation for training quality, safety, and environmental benefits. In addition, an increased use of scenario-based training will be incorporated into the curriculum.

Personal Aviation

Personal aviation refers to flights that are conducted for transportation or recreational purposes, without any financial remuneration. This mode of transportation is gaining momentum as an efficient method of commuting from one location to another.

Very light jets (VLJs) have created new personal aircraft options that fly faster, higher, and are more technologically advanced than previously available in GA. In many cases, the technology available on VLJs in more advanced than what is available on airline flights (such as synthetic-vision G1000 displays).

Very Light Jets (VLJs)

In recent years, GA aircraft have experienced a rapid increase in technological sophistication with the introduction of glass cockpit avionic displays. In addition, very light jets (VLJs) are set to revolutionize GA and the aviation industry as a whole. VLJs provide the speed and efficiency of jet-powered flight to moderately experienced pilots at a reasonable cost of ownership and are certified for single pilot operations. The first VLJs entered into active service in 2007. The FAA predicts that by 2020, 6,300 VLJs will be in operation (FAA 2007). This results in GA pilots flying faster and more technologically advanced aircraft than previously available.

A very light jet (VLJ) is defined as a jet aircraft weighing 10,000 pounds or less maximum certified takeoff weight and certified for single pilot operations. These aircraft will possess at least some of the following features: (1) advanced cockpit automation, such as moving map GPS and multifunction displays; (2) automated engine and systems management; and (3) integrated autoflight, autopilot and flight-guidance systems (NBAA Safety Committee 2005). Some of the additional characteristics of VLJs include:

- They cost less than one-half the price of the least expensive traditional jet (approximately $2 million dollars for a VLJ).

- They typically have five to six seats.

- They can land on runways much shorter than traditional jets, allowing operation out of underused GA airports

- They can be operated by a single pilot, rather than two-pilot crews, which are required for larger jets.

There are several manufacturers of VLJs, including Piper and Cirrus Aircraft. However, there is only one Canadian manufacturer of VLJs: Diamond Aircraft.

Figure 2.2 *The Diamond D-Jet, a Canadian-manufactured VLJ, in flight.*

Aircraft Manufacturing—Diamond Aircraft

Canada is a major developer of aerospace products. On an annual basis, $20 billion dollars worth of Canadian aerospace products are sold throughout the world. Some examples of Canadian companies that are leaders in the aerospace industry include Bombardier, CAE, Mechtronics, and Diamond Aircraft.

Diamond Aircraft began in Austria as Dimona Motor-glider. In 1993, CEO Christian Dries expanded Diamond Aircraft by purchasing a manufacturing facility in London, Ontario Canada. Since that time, Diamond Aircraft has established an international reputation for the style, efficiency, and safety of its aircraft (primarily used for flight training and personal aviation). In fact, Diamond aircraft boasts the best safety record in the industry and, subsequently, low insurance rates.

Diamond's most recent development efforts have been focused on the development, testing, and certification of its VLJ—the D-Jet. The development of a jet is a much more complicated endeavour than that of a piston-engine aircraft. In jet aircraft, pressurization, hydraulics, higher airspeeds, and decreased flight stability at high altitudes must be considered. The D-Jet differs from other VLJs because it is a single-engine aircraft, producing significantly better fuel efficiency than multi-engine VLJ designs.

Before Diamond Aircraft, or any manufacturer, can sell a new aircraft to the public, it must be certified by Transport Canada. The certification process is lengthy, taking a minimum of four years for a VLJ. During the certification process, a manufacturer must prove that its aircraft conforms to various standards. Diamond employs a complete ground test department which does nothing but extreme tests on the D-Jet. Examples of these tests include:

1. Adding extreme weight to the body of the aircraft to determine ultimate break points

2. Continually running a fuselage through pressurization and depressurization cycles to simulate fifty years of daily operation

3. Increasing the pressure on the fuselage to point where something breaks to see what breaks first

4. Complete flight testing program evaluating critical phases of flight, well beyond the limits of what any normal pilot would encounter

In certification, an aircraft is taken to the outer-extreme. Through this process, snags are identified and a manufacturer is forced to go back to the drawing board. This results in additional time and costs. However, the result is that the public can feel safe while flying in a certified aircraft, knowing that it has been tested far beyond normal flight characteristics!

The D-Jet is designed to be pilot-friendly and easy to fly, making it an excellent jet trainer and personal aviation jet. It is sold fully equipped with advanced glass-cockpit avionics, including the popular Garmin G1000. Diamond is positioned to deliver the first fifty D-Jets in 2010, 125 in 2011, and 200 a year from 2012 on. Through work on the D-Jet, Diamond is set to reshape the future of personal jet travel in Canada and worldwide.

PRELIMINARY SPECIFICATIONS

Cabin Seating:	2 pilots + 3 passengers
Maximum ramp weight:	5690 pounds
Useful load:	2240 pounds
Maximum fuel:	1740 pounds
Maximum cruise speed:	315 ktas
Certified ceiling:	25,000 feet
Long range cruise speed:	240 ktas
Maximum range:	1,350 nm
Cost:	1.4 million USD

Personal Aviation Challenges

Although there have been significant advances in personal aviation, COPA (2008) identifies several challenges facing this sector. These challenges include maintaining access to airports, fuel supply, and environmental issues.

One challenge facing the GA sector is maintaining access to airports for small personal aircraft. There are 730 certified airports in Canada and thousands of uncertified aerodromes. However, the airlines serve only eighty of these airports. Much of Canada (particularly in northern and rural areas) is accessible **only** by small aircraft. Often, these small airports represent the only delivery method of medicine and supplies. It is unlikely that roads or railways will ever be built to remote or arctic areas. Therefore, ensuring the viability of the network of smaller airports is crucial to many communities across the country.

A second challenge identified by COPA (2008) is the uncertainty over the availability and affordability of aviation fuel—specifically 100 Low Lead (100LL). Several types of alternative fuels are being researched, including:

- Biofuel: Derived from soybeans, this option may be the most practical alternative aviation fuel. However, additional research is being done to stabilize the fuel to provide the reliability required in aviation.

- Alcohols (such as ethanol): These alcohols are derived from corn, but they are heavier and take up more volume per unit of energy than traditional aviation fuel. This is a major issue in aviation as heavier fuel means less revenue-generating payload.

- Synthetic oil-based fuel (synfuel): This fuel has the same characteristics as traditional aviation fuel, but it is cleaner with less emissions.

- Hydrogen: A naturally-occurring element, hydrogen shows great potential as the aviation fuel of the future. However, there are still questions to be answered, including the storage, transportation, and safety of hydrogen aviation fuel. In addition, there is a negative stigma associated with hydrogen fuel because of the *Hindenburg* disaster.

A viable alternative fuel option is not yet available for the aviation industry. However, in upcoming years it can be anticipated that one of these fuel options will be perfected and implemented in aviation.

CASE — HINDENBURG DISASTER

In 1900, Ferdinand von Zeppelin, a German, was the first to fly a rigid airship. The structure of the ship was made of aluminum, and it was filled with hydrogen gas, which provided the lift, and propelled by two engines. In 1909, multiple airports were built across Germany, and DERLAG was formed as the world's first passenger airline. In the 1930s they created the largest airship in the world, which was named the *Hindenburg*. In this post-World War I era, the United States had all of the helium and refused to provide it to Germany because they feared it would be used for military applications. Therefore, the *Hindenburg* was filled with hydrogen gas, which provided good lifting characteristics but was extremely flammable.

On 6 May 1937, the *Hindenburg* was completing a journey from Frankfurt, Germany, to the Lakehurst Naval Air Station in New Jersey, United States. While attempting to dock at the Air Station, the *Hindenburg* caught fire. Due to the flammable nature of the hydrogen gas, it took less than one minute for the fire to completely destroy the airship. Thirty-six people were killed in the incident, while sixty-one were able to survive the crash.

The aviation industry worldwide contributes 3 percent of global emissions. However, even though there are more aircraft and hours flown in GA, GA contributes less than 1 percent (COPA 2008). This statistic includes all of the flight training that is conducted worldwide. Therefore, eliminating GA would have a tremendous negative impact on the economy and on transportation freedoms, with negligible environmental benefits.

There are significant challenges that face the future of personal aviation. However, COPA is working to represent the interests of personal aviation and ensure that this option remains open and available for Canadians.

Corporate Aviation

A unique sector of GA is corporate aviation. Corporate aviation refers to flight activities that are conducted in a corporately-owned aircraft. For example, if a large company owned a private jet to transport its executives to and from meetings, it would maintain a flight line. Most likely, the flight line will be operated by dispatchers, maintenance staff, and a team of pilots who are employed by the company. The team of pilots would rotate on-call duty, taking turns being ready to respond immediately if someone in the company requires transportation.

However, maintaining a flight line and the associated staff is too great an expense for many corporations. Therefore, a relatively new concept in corporate aviation is fractional ownership. Fractional ownership allows companies or individuals to purchase a portion of the aircraft. For example, a company pays one-eighth of the aircraft price to a fractional ownership operation. Then, whenever someone in the company needs transportation, he or she has access to the aircraft unless one of the other owners is already using it. In this situation, the fractional ownership operation will try to provide a similar aircraft in its fleet to meet the company's transportation needs.

Fractional ownership operations usually do not qualify under CARs 604 private operator rules, as a company usually maintains the operation and provides the pilots. Typically, these operations fall under CARs 703 air taxi operations and subsequently have some restrictions on operations and higher training standards than corporate flight departments.

There are significant advantages associated with fractional ownership, as executives do not have to wait in airport security and check-in lines, and the aircraft can land at a satellite airport which may be closer to the desired destination. However, the purchase cost of the aircraft is not the only fee required. Companies must also pay for each hour flown, to cover maintenance and fuel costs, and pay a monthly fee to the fractional ownership operation to cover storage and up-keep. Overall, for companies that charter aircraft between 100 and 400 hours a year, fractional ownership represents a cost-effective and convenient method of transportation (Careless n.d.). It is expected that fractional ownership will increase in popularity with the influx of affordable VLJs in upcoming years.

Helicopter Operations

There are a wide range of helicopter operations in Canada. The majority of flight hours occur in resource-related activities, including forestry, firefighting, mining, and geology-related activities. Beyond this, smaller components of the industry include heli-jets (scheduled helicopter services), flight training, off-shore work, business aviation, heli-tourism (sight-seeing), and emergency medical services (EMS).

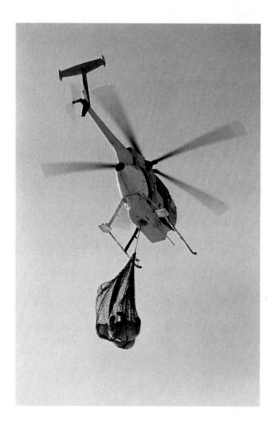

Photos.com.

Figure 2.3 *A helicopter carrying an external load.*

BUSH FLYING

Helicopter operations are chosen only when a task cannot be accomplished any other way, such as via road or float-plane transportation. The reason for this is that helicopter transportation is the most expensive option, at least three times the cost of comparable fixed-wing aircraft. Therefore, helicopter operations are often located in very remote areas that are not serviced by roads. This has several implications:

Work is typically completed only during the summer months, as winter weather makes year-round operations impossible. Pilots are hired on a contract-basis for four to five months.

- Pilots and crew live in bush camps. Although geographically remote, modern bush camps are outfitted with many amenities, such as television, wireless Internet, and separate cookhouses.

- The day-to-day activities of pilots often include transporting crew to a specific hard-to-reach location, dropping them off, and picking them up later in the day. Operations may also include long-line work where an external load (such as a drill or fuel drum) is connected to the belly of the helicopter via a long rope. The summer work schedule includes a lot of long hours, but many helicopter pilots find it very rewarding.

- Major maintenance is completed during winter months.

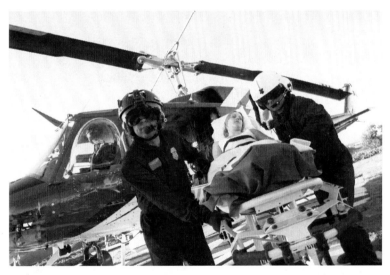

Shutterstock © Monkey Business Images.

Figure 2.4 *Helicopters are commonly used for medical evacuations, termed MEDEVAC.*

FIXED-WING VS. ROTARY-WING AVIATION

Fixed-wing aviation is a relatively predictable process that is well-supported by infrastructure. The aircraft takes off from a specific airport, is in contact with air traffic control en route, and lands at a destination airport with predictable services (such as fuel or maintenance). By comparison, rotary-wing aviation is highly unpredictable. A helicopter pilot often works with minimal support from air traffic control, particularly in remote areas, and never knows what to expect at a landing site. The landing area may have a helipad, or it might just be a field or confined area. The pilot must be trained to evaluate unusual situations and respond appropriately. Therefore, the benefits of experience are greater in a helicopter environment because of the unpredictable nature of the flying.

The fixed-wing and rotary-wing sections of the aviation industry are quite separate. Separate sections of Transport Canada deal with each. The skills required and working environment are significantly different.

Airlines

The creation of reliable scheduled air travel has reshaped how business is conducted, populations are distributed, and leisure time is spent. It is important to realize that the concept of travelling overseas for a weekend getaway was as foreign to our great-grandparents as science-fiction space travel is to the modern generation. Understanding that human beings have been in existence for approximately 2000 centuries, it is remarkable that within one century mankind was able to successfully achieve flight, break the sound barrier, and integrate air transportation into the mainstream.

Not all airlines are the same. Airlines can be categorized into several levels:

Major—operating passenger or cargo services with jet aircraft. Major airlines are segregated into legacy carriers, such as Air Canada, and low-cost carriers (LCCs), such as Westjet. The major airline industry in Canada is dominated by Air Canada, with Westjet being the only significant competitor. Jet operations employ approximately 40 percent of full-time pilots in Canada. Major operations are subject to CAR 705.

Regional—operating mostly twenty- to fifty-seat aircraft on scheduled passenger service. Examples of regional carriers include Air Canada Jazz, Bearskin and Porter. Regional operators employ approximately 16 percent of full-time pilots in Canada. Regional operations are subject to CAR 705.

CANADIAN AVIATION REGULATIONS (CARs) SECTION 705.01 THIS SUBPART APPLIES TO OPERATION BY A CANADIAN AIR OPERATOR, IN AN AIR TRANSPORT SERVICE OR IN AERIAL WORK INVOLVING SIGHTSEEING OPERATIONS, OF ANY OF THE FOLLOWING AIRCRAFT:

(A) AN AEROPLANE, OTHER THAN AN AEROPLANE AUTHORIZED TO OPERATE UNDER SUBPART 4, THAT HAS A MAXIMUM CERTIFIED TAKEOFF WEIGHT (MCTOW) OF MORE THAN 8618 KG (19,000 POUNDS) OR FOR WHICH A CANADIAN TYPE CERTIFICATE HAS BEEN ISSUED AUTHORIZING THE TRANSPORT OF 20 OR MORE PASSENGERS;

(B) A HELICOPTER THAT HAS A SEATING CONFIGURATION, EXCLUDING PILOT SEATS, OF 20 OR MORE; OR

(C) ANY AIRCRAFT THAT IS AUTHORIZED BY THE MINISTER TO BE OPERATED UNDER THIS SUBPART.

Commuter large aircraft—operating ten- to nineteen-seat turbo-prop aircraft on scheduled service. Large commuter operations employ approximately 13 percent of full-time pilots in Canada. Commuter operations are subject to CAR 704.

Commuter small aircraft—operating six- to nine-seat fixed-wing aircraft on scheduled service. Approximately 5 percent of full-time pilots in Canada are employed in this sector. Commuter operations are subject to CAR 704.

Examples of commuter operations in Canada include Air Georgian, Central Mountain Air, Calm Air, and Provincial Airlines.

> **CARs SECTION 704.01 APPLIES IN RESPECT OF THE OPERATION BY A CANADIAN AIR OPERATOR, IN AN AIR TRANSPORT SERVICE OR IN AERIAL WORK INVOLVING SIGHTSEEING OPERATIONS, OF ANY OF THE FOLLOWING AIRCRAFT:**
>
> **(A) A MULTI-ENGINED AEROPLANE THAT HAS A MCTOW OF 8618 KG (19,000 POUNDS) OR LESS AND A SEATING CONFIGURATION, EXCLUDING PILOT SEATS, OF 10 TO 19 INCLUSIVE;**
>
> **(B) A TURBO-JET-POWERED AEROPLANE THAT HAS A MAXIMUM ZERO FUEL WEIGHT OF 22680 KG (50,000 POUNDS) OR LESS AND FOR WHICH A CANADIAN TYPE CERTIFICATE HAS BEEN ISSUED AUTHORIZING THE TRANSPORT OF NOT MORE THAN 19 PASSENGERS;**
>
> **(B.1) A MULTI-ENGINED HELICOPTER WITH A SEATING CONFIGURATION, EXCLUDING PILOT SEATS, OF 10 TO 19 INCLUSIVE, UNLESS IT IS CERTIFIED FOR OPERATION WITH ONE PILOT AND OPERATED UNDER VFR; AND**
>
> **(C) ANY AIRCRAFT THAT IS AUTHORIZED BY THE MINISTER TO BE OPERATED UNDER THIS SUBPART.**

Air taxi—operating smaller aircraft on a "taxi" system between small airports, meaning that flights are on-demand and not scheduled in advance. This type of operation also includes air ambulance, fire suppression, and aerial work. Approximately 25 percent of full-time pilots in Canada are employed in air taxi operations. Air taxi operations are subject to CAR 703 (CAMC 2003).

> **CARs SECTION 703.01 APPLIES IN RESPECT OF THE OPERATION BY A CANADIAN AIR OPERATOR, IN AN AIR TRANSPORT SERVICE OR IN AERIAL WORK INVOLVING SIGHTSEEING OPERATIONS, OF ANY OF THE FOLLOWING AIRCRAFT:**
>
> **(A) A SINGLE-ENGINED AIRCRAFT**
>
> **(B) A MULTI-ENGINED AIRCRAFT, OTHER THAN A TURBO-JET-POWERED AEROPLANE, THAT HAS A MCTOW OF 8618 KG (19,000 POUNDS) OR LESS AND A SEATING CONFIGURATION, EXCLUDING PILOT SEATS, OF NINE OR LESS**
>
> **(B.1) A MULTI-ENGINED HELICOPTER CERTIFIED FOR OPERATION BY ONE PILOT AND OPERATED UNDER VFR; AND**
>
> **(C) ANY AIRCRAFT THAT IS AUTHORIZED BY THE MINISTER TO BE OPERATED UNDER THIS SUBPART**

Airline Management

Costs

As one might expect, managing and operating an airline is a very complicated endeavour. The airline industry has a reputation for very slim profit margins. The single greatest expense for most airlines is labour costs. Airlines must cover the salaries of a large number of employee groups, including pilots, flight attendants, dispatchers, maintenance staff, ticketing and gate agents, baggage handlers, ground support, and call-centre employees. Typically, these employees must be distributed throughout a large network of airports.

The second largest expense for an airline is fuel. When fuel prices increase, there is a direct and significant impact on airline profits. It has been estimated that a $1 increase in the price of fuel costs the airline industry $425 million (Alexander 2004). Another challenge the airlines face is the enormous debt associated with new aircraft, which are typically leased rather than purchased outright.

The weather is another factor which impacts airline profits, and is inherently uncontrollable. Once en route, airline dispatchers work with pilots to avoid bad weather. It doesn't matter how good an aircraft is built or how much experience a pilot has, a commercial flight cannot fly through a thunderstorm. However, when a severe weather system overlies an airport, there is little that can be done except wait for it to pass. The resulting grounding of aircraft and re-routing of passengers is tremendously expensive. Another expense associated with weather is de-icing, which is common during the Canadian winter. The central de-icing facility (CDF) at Toronto's Pearson International Airport has twenty-seven de-icing machines, each of which costs approximately $1 million (Morris 2009).

Revenue

Airlines make the majority of their money from passenger tickets. All tickets on an aircraft are not set at the same rate. The process of assigning prices to tickets is extremely complicated and is accomplished through computerized systems. Ticket prices vary, based on the days until departure, forecasted demand, day and time of departure, and load factor. The load factor is the percentage of available seats filled by passengers. For example, if a 200-passenger aircraft completed a trip with only 100 passengers, the load factor is 50 percent. This is an important statistic because airlines calculate a break-even load factor, meaning that load factors less than a certain percentage will result in flights costing the airline money! After the devastating events of 9/11, airlines were regularly operating below break-even load factors—losing money on every flight!

In addition to load factors, an important term in airline economics is the *yield*. The *yield* refers to the revenue per seat mile (RSM). For example, if a passenger is flying a 1000-mile trip and the passenger's fare was $100, it would result in a yield of 10 cents. It is important that both the load factor and the yield be considered by an airline, as they must exist in a balance. Discount tickets may result in high load factors but low yields, resulting in an airline losing money. On the other hand, very expensive tickets would result in high yields but fewer passengers and a lower load factor. This would also decrease the airline's ability to make money. Balancing the airline's yield and load factor represents one of the major challenges in airline management.

However, passenger tickets are not the only source of airline revenue. It is also common for airlines to carry freight. Some airlines, called *cargo airlines*, carry freight exclusively. Examples of these companies include United Parcel Service (UPS) and Federal Express. Onboard amenities, including food and drink purchases, and advertising are additional sources of revenue.

Airline Scheduling

Scheduling is an important, yet complicated, aspect of airline management. There are three main components of the airline scheduling process: schedule, aircraft routing, and crew pairing.

Schedule

Schedulers must determine which airports to service on which days and at which times. In addition, schedulers will consider the arrival times of feeder flights for connection passengers, flight time between airports, and turnaround times on the ground. The schedule must be created with consideration for the available landing slots at airports, meeting on-time goals, and avoiding delays in the network. The delay of one flight has tremendous ramifications throughout the airline's network of activities. A delay "chain" exists. Delays which occur early in the day impact the entire network. The result is like dropping a pebble in a pond—the ripples expand and cross the entire body of water. The reasons for this impact include passengers missing connections and requiring re-routing, the subsequent leg scheduled for the aircraft may be delayed, the crew scheduled for the aircraft may exceed duty-time limitations.

An approach that is used by most airlines is the hub-and-spoke system. In a wheel, the hub is in the centre, and many spokes extend out in different directions. In an airline, a "hub" airport is identified. Air Canada's hub airports are Montreal, Toronto, and Vancouver. From the airline's hub airport, routes will be developed to spoke airports around the country. There are several advantages of the hub-and-spoke system. The greatest advantage is that the airline can offer service to small spoke airports that might not otherwise have air service. This system also allows major airlines to have full-time maintenance and support staff at its hubs, as a large number of personnel are required to assist with the flow of passengers and baggage to connecting flights. In addition, centralizing operations allows for improved customer service and rapid turnaround times on the ground.

However, there are disadvantages associated with the hub-and-spoke system. Due to the high volume of traffic arriving and departing at hub airports, air traffic congestion is common and delays often occur during peak times. Travelling through a hub-and-spoke system also means that passengers are inconvenienced by making several connections, transferring through hub airports.

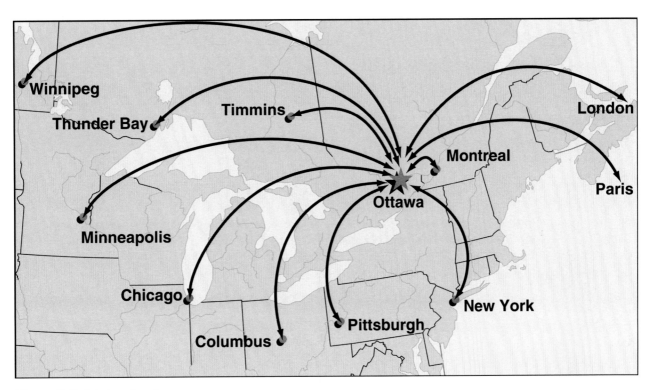

Figure 2.5 *A hub-and-spoke system.*

The lesser-utilized alternative to the hub-and-spoke system is a point-to-point approach. Rather than operating flights out of a hub airport, airlines that use a point-to-point approach do not have a hub airport and may offer service from several key markets. For example, Westjet's point-to-point system operates international flights out of London, Hamilton, and Ottawa. On Air Canada's hub-and-spoke system, a passenger in one of these cities would have to connect in Toronto before continuing on to their international destination. The benefit of a point-to-point approach is that passengers have to make fewer connections, as they can get a direct flight without transferring through a hub airport. In this manner, point-to-point airlines are connecting spoke airports. The disadvantage of this approach is that the number of flights is limited to heavily-travelled destinations. Therefore, the point-to-point market is highly effective for travel to tourist destinations like Las Vegas or Orlando. However, it may be too limited for business travelers who wish to travel anywhere in the country.

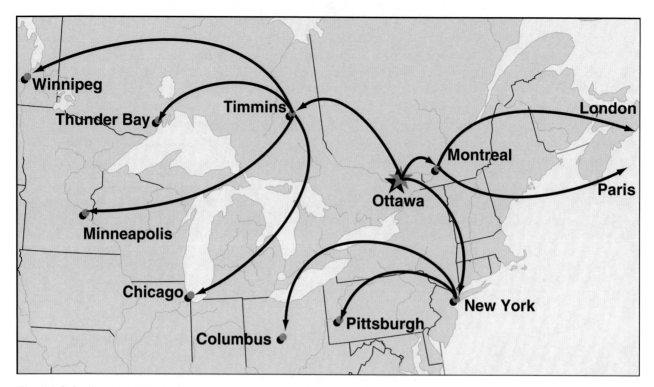

Figure 2.6 *A point-to-point system.*

Aircraft Routing

Due to the high cost of transport aircraft, airlines do not keep reserve aircraft sitting around at the airport. Each aircraft in an airline's fleet is utilized to its maximum capacity, with due consideration being given to required maintenance delays. Schedulers must determine how to best distribute the fleet throughout the schedule by considering the passenger capacity, range of aircraft, and curfew restrictions of airports. Some airports have noise restrictions that prohibit takeoffs and landings after curfew. These restrictions have major implications on aircraft scheduling and are the reason why a lot of overseas flights occur at night.

Crew Pairing

Organizing and scheduling crews for each flight are another important aspect of scheduling. Every flight must be staffed with pilots and flight attendants. However, crew pairing is a very complicated issue because pilots have maximum duty times. By regulation, pilots are not allowed to work past maximum duty times. The reason for duty-time limitations is, in the interest of safety, to avoid unsafe amounts of flying and ensure that pilots get adequate rest between legs.

CAR 720.15 FLIGHT TIME LIMITATIONS STATES THAT "FOR ANY 6 NON-OVERLAPPING PERIODS OF 30 CONSECUTIVE DAYS WITHIN A 365 CONSECUTIVE DAY PERIOD, THE MAXIMUM FLIGHT TIME IN ANY AIRCRAFT SHALL NOT EXCEED:

(A) WHERE THE FLIGHT CREW MEMBER CONDUCTS SINGLE-PILOT IFR OPERATIONS, 8 HOURS IN ANY 24 CONSECUTIVE HOURS;

(B) 60 HOURS IN ANY 7 CONSECUTIVE DAYS;

(C) 150 HOURS IN ANY 30 CONSECUTIVE DAYS

(D) 210 HOURS IN ANY 42 CONSECUTIVE DAYS;

(E) 450 HOURS IN ANY 90 CONSECUTIVE DAYS;

(F) 900 HOURS IN ANY 180 CONSECUTIVE DAYS;

(G) THE ACCUMULATED 30-CONSECUTIVE DAY, 42-CONSECUTIVE DAY AND 90-CONSECUTIVE DAY FLIGHT TIMES MAY BE RESET TO ZERO IF THE FLIGHT CREW MEMBER IS PROVIDED WITH AT LEAST 5 CONSECUTIVE DAYS FREE FROM ALL DUTY; AND

(H) 1200 HOURS IN ANY 365 CONSECUTIVE DAYS.

CAR 720.16 FLIGHT DUTY TIME LIMITATIONS AND REST PERIODS STATES THAT "THE MAXIMUM FLIGHT DUTY TIME MAY BE EXTENDED TO 15 CONSECUTIVE HOURS IF:

(A) THE MINIMUM REST PERIOD IS INCREASED BY 1 HOUR; OR

(B) THE MAXIMUM FLIGHT TIME DOES NOT EXCEED 8 HOURS IN ANY CONSECUTIVE 24 HOURS."

WHEN "THE FLIGHT CREW IS AUGMENTED BY THE ADDITION OF AT LEAST ONE FULLY QUALIFIED FLIGHT CREW MEMBER, FLIGHT DUTY TIME MAY BE EXTENDED TO 15 CONSECUTIVE HOURS IF:

(A) THE ADDITIONAL FLIGHT CREW MEMBER OCCUPIES A FLIGHT DECK OBSERVER SEAT DURING TAKE-OFFS AND LANDINGS UNLESS THE OBSERVER SEAT IS REQUIRED BY AN AIR CARRIER INSPECTOR, IN WHICH CASE, A PASSENGER SEAT MUST BE AVAILABLE FOR THE FLIGHT CREW MEMBER; AND

(B) THE SUBSEQUENT MINIMUM REST PERIOD IS INCREASED BY AT LEAST 12 HOURS."

(C) "WHERE THE FLIGHT CREW IS AUGMENTED BY THE ADDITION OF AT LEAST ONE FLIGHT CREW MEMBER, THE DIVISION OF DUTY AND REST IS BALANCED BETWEEN THE FLIGHT CREW MEMBERS AND A FLIGHT RELIEF FACILITY IS PROVIDED, FLIGHT DUTY TIME MAY BE EXTENDED IF:

(D) WHERE A FLIGHT RELIEF FACILITY-SEAT IS PROVIDED, THE FLIGHT DUTY TIME MAY BE EXTENDED TO 17 CONSECUTIVE HOURS, IN WHICH CASE THE MAXIMUM FLIGHT DECK DUTY TIME FOR ANY FLIGHT CREW MEMBER SHALL BE 12 HOURS;

(E) WHERE A FLIGHT RELIEF FACILITY-BUNK IS PROVIDED THE FLIGHT DUTY TIME MAY BE EXTENDED TO 20 CONSECUTIVE HOURS, IN WHICH CASE THE MAXIMUM FLIGHT DECK DUTY TIME FOR ANY FLIGHT CREW MEMBER SHALL BE 12 HOURS;

(F) THE SUBSEQUENT MINIMUM REST PERIOD SHALL BE AT LEAST EQUAL TO THE LENGTH OF THE PRECEDING FLIGHT DUTY TIME; AND

(G) A MAXIMUM OF 3 SECTORS MAY BE COMPLETED."

CANADIAN AIRLINES

A SUCCESS STORY

WestJet is a regional airline that was created in 1994. Westjet is a low-cost carrier (LCC) modeled after Southwest Airlines which operates out of the United States. Westjet started in Western Canada with only three aircraft and 247 employees. The core goals of Westjet have always been to provide excellent customer service at an affordable price. This is accomplished through efficient business models and a strong corporate culture, supporting employees so that they can take excellent care of customers. Westjet employees are also Westjet owners, through stock-purchasing and profit sharing plans, which encourages employees to contribute to the company's success on a daily basis. For example, job descriptions are flexible at Westjet, and everyone helps out. It's not unusual to see a pilot help board the passengers or flight attendants stick around to help clean the aircraft. All of the employees pitch-in to get the job done, which results in fast turnaround times and high aircraft utilization.

Some examples of Westjet's efficient business models include:

*   Using a point-to-point model, rather than a hub-and-spoke approach

*   Aircraft maintenance and pilot training costs are reduced because only one type of aircraft is used, the Boeing 737

*   Reducing labour costs, which are the single greatest expense of an airline, by paying employees less and offering a profit-sharing model to retain staff

Since its creation, Westjet has experienced explosive growth. From its creation in 1994 with only three aircraft, Westjet now has seventy-six Boeing 737 aircraft and 7,500 employees. In Canada, Westjet is the only significant competitor to Air Canada. Westjet has become the most profitable airline in North America and the fourth most profitable airline in the world! By 2016, Westjet's vision is to become one of the five most successful international airlines (*Financial Post* 2009).

A TRAGIC FAILURE

JetsGO was a LCC that started operations in 2002 with three aircraft. From the time it was created, JetsGO operated in direct competition with Westjet. At times, to win the competitive edge over Westjet, the airline offered airfares as low as $1. Yet all was not well at the doomed airline as it built a history of mechanical breakdowns, poor maintenance, low-time pilots, and in-flight mistakes. The focus of JetsGO management was on rapidly increasing fleet size and routes. Unfortunately, this growth came at the cost of sound operational practices and a safety program.

By 2005, JetsGO had managed to capture 10 percent of the domestic market in Canada. The JetsGO fleet had grown from three to twenty-nine aircraft. Unfortunately, in January of 2005, a JetsGO aircraft landing at the Calgary airport in poor weather slid off the runway, rolling across the grass and hitting a runway sign before taking off again. When Transport Canada investigated the incident, they found enough evidence to put special restrictions on JetsGO, even before completing the investigation. On March 8 of 2005, Transport Canada informed JetsGO management that their operating certificate would be suspended on the 9th of April unless problems with their poor organizational structure and flight safety program were corrected (Cribb, Vallance-Jones, and McMahon 2006). On 11 March 2005, JetsGO unexpectedly ceased all operations, leaving an estimated 17,000 passengers stranded and 1,200 employees out of work. JetsGO declared bankruptcy shortly thereafter.

Airline Employees

Pilot

To be hired as a pilot at a major airline, typically, a person is required to have an airline transport pilot's license (ATPL) with a valid medical certificate and a minimum of 1000 hours of flight experience. However, to be competitive in the hiring process, pilots require much more experience.

Entry-level pilot salary at a major airline is approximately $40,000/year. However, salary increases as pilots transition to larger aircraft, longer routes, and from the co-pilot's seat on the right-hand side of the cockpit to the captain's seat on the left. Senior captains who fly the largest aircraft on transoceanic routes make over $200,000/year. The age at which pilots are hired by a major airline ranges from the mid-20s to the late-40s. However, the average age of new airline pilots is in the mid-30s.

It is important to understand that pilots are supported by powerful unions. The Air Canada Pilot's Association (ACPA) is the bargaining unit that represents Air Canada pilots. Also active in Canada is the Air Line Pilots Association (ALPA) which is an international union that represents over 50,000 pilots around the world.

In a typical month, airline pilots complete eighty flight hours and work an average of fifteen days. However, an average of 300 hours is spent away from home (because some off-duty time is spent in other cities and countries). The amount of time away from home can make family life very difficult for pilots, at least until they have enough seniority to bid a more favourable schedule.

Pilot work schedules are based on a competitive bidding process. When a new pilot is hired at an airline, he or she is assigned a seniority number. Pilots' seniority numbers will be extremely important throughout their entire careers at a company. Pilots with the highest seniority numbers are given first priority to choose their flight schedule each month and vacation time. On the other hand, new pilots with low seniority numbers will have to fly the least desirable routes and will often work during holidays. However, it is possible to gain relative seniority on a certain aircraft type or by status (as a captain or first officer). For example, each time a pilot moves to a more senior plane or position, his or her seniority number will be low compared to other pilots flying that aircraft in that position. Therefore, if a pilot's schedule is very important to a person (perhaps because he or she has small children and wants to be home regularly) the pilot may choose to stay at the first officer level to gain relative seniority. On the other hand, if a pilot is interested in a more swift career progression, he or she may choose to upgrade whenever possible. This will result in a bigger paycheck but a relatively low seniority with an unpredictable schedule.

The reality of the airline industry is that it is heavily impacted by economic recessions. The incidents of 11 September 2001 and the economic crisis that began in the fall of 2008 have had devastating effects on airlines. During these difficult times, airlines often "furlough" pilots. A furlough is used to avoid layoffs. When a pilot is furloughed, he or she will not complete any work for the company and will not be paid. However, he or she will maintain his or her seniority number. When hiring begins again, seniority numbers will determine the order in which pilots are brought back into active service. All pilots will be brought off of furlough before new pilots are hired by the airline. Despite economic recessions, the demand for pilots is expected to increase steadily in coming years. In fact, by 2025 a significant pilot shortage is expected worldwide.

Figure 2.7 *The "office" of an airline pilot.*

Dispatchers

Dispatchers play a very important role within an airline. Dispatchers are well-trained and qualified personnel who are jointly responsible, with the captain of the aircraft they guide, for the safety and operational efficiency of flights. Second only to pilots, dispatchers are the second-highest paid group at an airline (entry-level positions start around $20,000/year and senior dispatchers at major airlines earn $100,000–150,000/year). The role of a dispatcher is to:

- Guide flights expeditiously based on company and government regulations

- Analyze weather maps to plan the most efficient routes and avoid hazards

- Compute fuel requirements

- Prepare flight plans

- Cancel or delay flights if unsafe conditions are identified

- Monitor weather and navigation charts in conjunction with aircraft position reports throughout the duration of flights

- Inform the flight crew of any changes that become necessary based on weather changes and recommends a new course or altitude (ADF 2009)

The workspace of a dispatcher is similar to that of an IFR air traffic controller. Dispatchers sit at a desk with several computer screens displaying the real-time location of aircraft they are guiding, as well as weather and flight planning information. Dispatchers play an important role in an airline, continually working to maximize route efficiency and safety.

Aircraft Maintenance

The importance of maintenance personnel within an airline cannot be overstated. Aviation maintenance professionals are highly-specialized and skilled individuals who are crucial to the success of an airline. There are over 14,000 Aircraft Maintenance Engineers (AMEs) in Canada. AMEs are licensed by Transport Canada to inspect, certify, and work on aircraft. AMEs earn a salary between $45,000 and 80,000/year.

There are several types of maintenance trades that work with an airline, including line maintenance, engine overhaul, avionics, sheet metal and plumbing, electrical, fabrication, machining, aircraft instruments, wheel and rubber units, among others.

All aircraft are required to complete maintenance inspections at intervals predetermined by Transport Canada. It is the responsibility of the maintenance crew to ensure that these inspections are completed efficiently and thoroughly.

Flight Attendants

Flight attendants are important airline personnel who provide customer safety service to passengers within the aircraft cabin. Flight attendant new-hires do not typically have any aviation background and complete a two-month training course at an airline training centre. During training, flight attendants will study cabin safety, security, *Canadian Aviation Regulations*, company policy and procedures, and customer service. Flight attendants must also be well-groomed.

Flight attendants earn an average salary of approximately $50,000/year. The role of a flight attendant can be challenging at times, because they are required to provide excellent customer service while ensuring the safety of the flight. This creates a unique authority dynamic whereby flight attendants must be assertive enough to encourage compliance (for example, when asking passengers to pay attention to security briefings or turn off cellular phones)—while still remaining pleasant and friendly.

On a recent United Airlines flight, a flight attendant had to physically body-check a disturbed passenger who was running towards the cockpit door. In the struggle that followed, while flight attendants and passengers were working to restrain the disturbed passenger, a passenger who was observing the incident asked the flight attendant when she would be able to bring him some more coffee. This is evidence that the conflicting security and customer service roles of flight attendants are not well-understood by passengers.

Once on the job, flight attendants will bid for their work schedule on a monthly basis. This process is based on a seniority number, similar to the pilot's schedule bidding process, except that flight attendants can fly on any type aircraft. The more senior the flight attendant, the more likely it is that he or she will be able to get the most desirable work schedule.

Conclusion

Civil aviation is an important industry within the fabric of the Canadian economy. The types of operations that are conducted within civil aviation are very diverse, including flight training, helicopter operations, personal aviation, and airlines. Some commercial operators thrive in Canada's northern wilderness, others succeed in transporting international jetsetters in luxurious private aircraft, and others provide one-of-a-kind sightseeing experiences over Niagara Falls. Whatever the operation, it is clear that Canada's civil aviation industry is a world-leader in promoting safe and efficient air transportation.

PRACTICE ACTIVITY

A wealthy investor has chosen you to create a new civil aviation operation. It is up to you to decide what type of operation you want to create and how you want to manage it. Complete the descriptions below to outline your ideas.

1. Name the type of operation and describe its primary activities: (such as flight training, helicopter, fractional ownership, air taxi, charter, cargo, airline).

    _____

    _____

    _____

    _____

    _____

2. Describe the type(s) of aircraft you will be operating and why they were chosen:

    _____

    _____

    _____

3. What staff do you require? What qualifications must they have? (How much experience will your pilots require? Will you need in-house maintenance, flight attendants, dispatchers?)

    _____

    _____

    _____

    _____

    _____

4. Outline why you believe that your operation will succeed in Canada. Is your concept new or has it already been established?

    _____

    _____

    _____

    _____

    _____

    _____

    _____

    _____

CHAPTER **3**

MILITARY AVIATION

OBJECTIVES

After completing this chapter, you will:

- Understand the modern history of military aviation in Canada

- Know the roles of Canada's Air Force, domestically and internationally

- Be familiar with the organization and rank structure of Canada's Air Force

- Know several non-commissioned officer and officer careers within the Air Force, including the selection, training, and lifestyle of Air Force pilots

- Understand and respect the importance of military heritage

Introduction

"Per ardua ad astra"—"through adversity to the stars"
Motto of Canada's Air Force

In 2009, Canada's Air Force celebrates its eighty-fifth anniversary. Throughout its existence, the Air Force has had a tremendous impact on world history through defensive, peacekeeping, and humanitarian efforts. This work has established an international reputation of Canadian military as a well-trained force that works to mitigate the effects of conflict. In fact, the idea of a United Nation's (UN) Peacekeeping Force was introduced by Lester Pearson, a Canadian Ambassador to the UN, in 1956 (Canada's Air Force 2008). As Chapter 1 reviews the history of aviation, including that of Canada's Air Force, this chapter will present information on the modern history and organization of Canada's Air Force.

Canadian Forces Photo.

Figure 3.1

Modern History of Canada's Air Force

In 1964, Lester B. Pearson's government proposed Bill C-90 Act to Amend the National Defence Act, which suggested the integration of the administrative structure of the Canadian Forces under one Chief of Defence Staff (CDS). The argument was made that the traditional organization of the three branches of the military with their own service chief and command structure was not conducive to a unified defence force which Canada required to meet the challenges of the future. In addition, the government was concerned that military spending was out of control and that unifying the military would eliminate unnecessary redundancies in administrative services (such as military police, medical, and postal services). Although there were some reservations within the Royal Canadian Navy, the changes were implemented (Lund 2007). In 1966, Bill C-243, Canadian Forces Reorganization Act, was proposed to unify the three branches of the Canadian military—the Canadian Army, Royal Canadian Navy, and Royal Canadian Air Force—and implement a common uniform and rank structure. The Royal Canadian Navy was so opposed to unification that its senior operational commander, Rear Admiral W. M. Landymore, was fired (Lund 2007).

Canada is the only member of the Commonwealth to unify its armed forces, which came into effect on 1 February 1968. At that time, the three branches of the military were unified into the Armed Forces. One of the most visible changes associated with unification was the standard green uniforms of all branches of the Armed Forces, which replaced the Air Force's blue uniforms that represented the sky above.

Unification was not a success, primarily because there was a lack of guidance from the government and resistance within the "old" services (Lund 2007). The financial savings that were expected from unification never materialized and equipment grew old and out-of-date. Although the Navy was the branch that was most opposed to unification, the Air Force was left with the fewest resources spread across five commands: Maritime (Navy), Mobile (Army), Air Defence, Air Transport, and Training (Watt 2009). Lieutenant-General Bill Carr was the first commander of Air Command, and he recognized the need for change. He began the process of documenting the underutilization of the air fleet and recommended that a formal air element of the Canadian Forces be reinstated (Watt 2009).

In 1975, the shift towards a traditional Air Force culture began. Air Command was created to restore structure to Air Force activities, including the former Army Flying Corps and Naval Fleet Air Arm. In 1986 the Air Force returned to its traditional light blue uniforms, and service chiefs were reinstated.

In 1997, the Air Force consolidated its aviation groups under a single command and control group called 1 Canadian Air Division. The Commander of 1 Canadian Air Division was given the Chief of Air Staff title and was moved to the National Defence Headquarters in Ottawa, Ontario (Watt 2009).

In 2005, the CDS General R.J. Hillier declared an initiative to reintroduce a joint force management structure in the Canadian Forces. The goal was for the forces to become more effective, integrated, and streamlined (Lund 2007). Currently, the Air Force provides a multipurpose, operationally ready, and combat capable force with the following roles:

- Domestic and international operations

- Airspace surveillance and control

- Search and rescue

- Humanitarian aid

Domestic and International Operations

The Canadian Forces has a deployable force of approximately 24,000 people. At any point in time, around 8,000 of those men and women are getting ready for, actively involved in, or returning from a mission overseas. There are several Commands in the Canadian Forces that have responsibility over different aspects of operations. This organizational structure was established in 2006. The Commands include:

- The Canadian Expeditionary Force Command (CEFCOM) which has responsibility for Canadian Forces operations outside of North America. CEFCOM deploys forces across the globe for humanitarian, peacekeeping, and combat operations. The task forces deployed by CEFCOM may include members of the Navy, Air Force, Army, or Special Forces (CEFCOM 2009).

- Canadian Special Operations Forces Command (CANSOFCOM) is the agency that provides the CDS and operational commanders with skilled and ready special operations forces (SOF) who conduct operations domestically and abroad. SOFs are utilized in special operations when traditional CF actions may not suit the task. CANSOFCOM is composed of the Joint Task Force 2, Canadian Special Operations Regiment, 427 Special Operations Aviation Squadron, and the Canadian Joint Incident Response Unit—Chemical, Biological, Radiological, and Nuclear. CANSOFCOM's core tasks are (1) counter-terrorism, (2) maritime counter-terrorism, and (3) high value tasks (which may include special reconnaissance, defence, diplomacy and development, and non-combatant evacuation operations) (National Defence 2008).

- Canada Command (Canada COM) is responsible for all routine and emergency CF operations in Canada and continental North America. The purpose of Canada COM is to provide: (1) a single point of contact for civil authorities seeking CF assistance, (2) a specific military command for North American operations, and (3) a focus on Canada as a distinct theatre of operations (National Defence 2009).

- Canadian Operational Support Command (CANOSCOM) provides national-level support to domestic and international CF missions. Examples of support activities include logistics, military engineering, health services, communications, and military police. CANOSCOM also plans, coordinates, and executes operational support for theatre activation, sustainment, and termination (National Defence 2006).

Airspace Surveillance and Control

North American Aerospace Defence Command (NORAD)

The North American Aerospace Defence Command (NORAD) was formed as a partnership between Canada and the United States to provide defensive intelligence. The purpose of NORAD is to control and provide warning (detection, validation, and warning of attack) of airspace overlying North America. The commander of NORAD is appointed jointly by the prime minister of Canada and the president of the United States, and is responsible to both parties. The commander of NORAD is located at the NORAD–U.S. Northern Command at Peterson Air Force Base in Colorado, United States. From this location, worldwide sensors are utilized to accurately identify threats and communicate this information to the Canadian and American leaders.

Three subordinate NORAD locations are at Canadian Forces Base at Winnipeg, Manitoba; Elmendorf Air Force Base, Alaska; and Tyndall Air Force Base, Florida. To accurately identify threats, NORAD uses an assortment of ground-based radar stations, satellites, airborne radar, and fighter aircraft which can intercept and engage threats if necessary. NORAD also assists in the identification of aircraft suspected of trafficking drugs into North America illegally (North American Aerospace Defense.Command n.d.).

Peacekeeping

Peacekeeping is an important task, but it is difficult to define. The 2nd United Nations (UN) Secretary-General described peacekeeping as falling between two chapters of the UN Charter. Peacekeeping falls between Chapter 6, which describes negotiation and mediation as peaceful methods of resolving disputes, and Chapter 7, which authorizes forceful action. The specific missions of peacekeeping involve maintaining ceasefires, stabilizing situations, human rights monitoring, disarmament, and reintegration of combatants (United Nations Peacekeeping 2008).

The Canadian Air Force is a major contributor to UN peacekeeping efforts, flying desperately needed equipment and people to countries in turmoil (primarily in Asia, the Middle East, and Africa). In addition, the Air Force provides patrol, fighter, and airlift services (Fetterly 2006).

However, the reality is that there is a price associated with peacekeeping work. In 2005–6 the Department of National Defence (DND) received a budget of $14.3 billion, of which $927 million was spent on peacekeeping activities through wages, fuel, equipment, and parts (Fetterly 2006). The debate over this issue is based on the foundational principle that the priority of the DND is to defend Canadian interests and values, while peacekeeping on an international scale is a secondary priority. Ultimately, Canada's Armed Forces have established an international peacekeeping reputation which is a source of pride to all Canadians. Although peacekeeping activities may not directly defend Canadian interests, many argue that preserving peace does defend Canadian values on an international scale and is therefore an important DND responsibility.

Search and Rescue

The National Search and Rescue Secretariat (NSS) is an independent government agency that reports to the Minister of National Defence. The purpose of the NSS is to promote the National Search and Rescue (SAR) program to facilitate effective and economical SAR programs across the country.

Although the military plays a major role in SAR activities, they do not work in isolation. In fact, the role of NSS is to coordinate the federal SAR activities of the Canadian Forces, Canadian Coast Guard, Royal Canadian Mounted Police, Transport Canada, Meteorological Service of Canada, and Parks Canada. The NSS also coordinates the non-federal agencies who contribute to SAR activities, including police services across the provinces and territories (National Search and Rescue Secretariat 2009).

Within the Air Force, more than 700 people are dedicated to SAR activities. The geographic make-up of Canada makes a highly skilled SAR team necessary. Canada's coastline is the world's longest, and the land mass in Canada is the second largest worldwide. Due to these vast distances, Canadian SAR helicopters must fly with bigger crews, more equipment, and larger fuel loads than SAR activities in other countries. An additional challenge associated with Canadian SAR activities are the wide variations in terrain and climate as teams

http://www.airforce.forces.gc.casiteathomedocimagessarmap2.gif.

Figure 3.2 *By comparing the size of Europe to Canada, it is clear that the size of Canada presents SAR challenges.*

must be able to operate in both summer and arctic conditions (Canada's Air Force 2007). The large landmass and coastline of Canada is evident in Figure 3.2, which compares Canada to the entire continent of Europe.

The purpose of a SAR team is to search, provide medical aid, and transport survivors. The day-to-day reality of SAR professionals is that they often risk their lives to save others, which is why their motto is "That others may live" (Canada's Air Force 2007). The Canadian Forces SAR team is considered to be one of the best in the world, trained to work as a team and deal with climate and terrain extremes.

CASE

THE SEARCH AND RESCUE OF BOX TOP 22

By Simon John Levesque

Simon John Levesque.

Figure 3.3

The Canadian Search and Rescue operation has played a vital role in aviation history, saving countless souls since its creation to this very day. One of the most famous SAR operations was the search and rescue of Box Top Flight 22 on 30 October 1991. Box Top Flight 22 was a Canadian Forces CC-130 Hercules transport aircraft tasked with bringing fuel, supplies, and personnel to Canadian Forces Station Alert, a listening post located just south of the North Pole. With the airstrip in view, the pilot made a fatal error, and the Hercules crashed only a few miles south of

continues

THE SEARCH AND RESCUE
OF BOX TOP 22 (CONTINUED)

the station. Out of the eighteen personnel on board, thirteen survived the crash and had to face the brutal cold from a Canadian arctic storm that greatly reduced their chances of surviving through the night. Their only chance to survive was to use the remaining tail section of the Hercules as a shelter and wait for rescue. The Canadian Search and Rescue Units were in a race against time to reach the survivors before they succumbed to their injuries and perished from exposure to the harsh and unforgiving climate of the Canadian high Arctic. However, the Crash of Box Top 22 was the first MAJAID ever to confront Canadian Forces Search and Rescue. A MAJAID is a "major air disaster" designated to an aircraft accident with ten or more casualties. In response to the MAJAID, the Joint Rescue Command Centre coordinated the largest search in Canadian history, mobilizing rescue aircraft from across the country and ordering them to Alert as fast as possible. Attempts to reach the survivors were also made by SAR-Techs in Snow-Cats by driving from CFS Alert over the ice to reach the crash site. However, white-out conditions from the arctic blizzard prevented them from reaching the survivors until their third attempt. The Hercules aircraft arrived on scene first, and SAR Techs parachuted down

to the survivors amid forty-knot winds, reduced visibility, and the perpetual night. They arrived just in time to treat the survivors with first aid, giving them enough time until the rescue helicopters arrived and evacuated all thirteen survivors to CFS Alert for treatment. However, the long journey to Alert from the Search and Rescue Squadrons in Edmonton, ALB; Trenton, ON; Greenwood, NS; and Gander, NL, well-exceeded the range of the Labrador Helicopters. Out of all the Labradors that were ordered to Alert, all were turned back due to mechanical failure except Rescue 315 piloted by Captain Mark Levesque and Captain Pierre Bolduc, who flew their Labrador nonstop from Trenton to Alert, a distance of over 5000 kilometres in 97.9 hours. Robert M. Lee, author of *Death and Deliverance*, commented that "their remarkable dash to the Pole was the stuff of aviation legend," because never before had a Labrador been flown so far, for so long, in such harsh conditions (Lee 1992). The rescue of the survivors of Box Top 22 was a complete success, and the brave men and women of the Canadian Search and Rescue Units who risked their lives to save the lives of their fellow countrymen, embodied the motto and spirit of Search and Rescue, "That others may live."

Simon John Levesque.

Figure 3.4 *Captian Mark Levesque in front of a Labrador helicopter.*

62   *Military Aviation*

Humanitarian Aid

The Air Force also contributes humanitarian aid around the globe. Since 1990, humanitarian missions have taken the Canadian Forces to Rwanda, Haiti, Honduras, and Turkey. In 1994, after two Field Ambulance teams arrived in Rwanda, too late, after the peak of cholera outbreak had passed, the Canadian government identified the need for a rapid-response humanitarian team (National Defence 2005). The incident in Rwanda led to the creation of the Canadian Forces Disaster Assistance Response Team (DART).

Disaster Assistance Response Team (DART)

The purpose of DART is to allow rapid deployment of approximately 200 military personnel for up to forty days to areas of the world experiencing natural disasters or humanitarian emergencies. DART is designed to deploy only to parts of the world where no military resistance or threats exist. DART will be deployed based on a request from the UN or a specific country, although the decision to deploy rests with the Canadian government. Specifically, DART serves four needs:

- "Primary medical care

- Production of safe drinking water

- A limited specialist engineer capability

- A command and control structure that allows for effective communications between the DART, the host nation, and the other agencies involved in the relief effort, including international organizations, non-governmental organizations and UN aid agencies" (National Defence 2005)

The Air Force was able to improve its international reach for humanitarian aid with the addition of four CC-177 Globemaster III aircraft in 2008.

CF Photo by Master Corporal Robert Bottrill.

Figure 3.5 *A CC-177 Globemaster III strategic transport aircraft from 429 Transport Squadron in Trnton, Ontario, rests on the airstrip at Kandahar Airfield.*

CAF Organization
Wings and Squadrons

The Canadian Air Force is made up of thirteen Wings across the country. *Wing* is a term used to describe a group of various operational and support units, led by a single tactical commander who reports to the operational commander. Each Wing is different, ranging in operations, purpose, and size (from a few hundred personnel to a few thousand) (Canada's Air Force 2006).

Canada's Air Force, 2006.

Figure 3.6

The following is the Air Force description of several of the thirteen Wings across Canada (Canada's Air Force 2006):

- **1 Wing Kingston** is the home of the Griffon helicopter. It provides airlift support of troops and equipment anywhere in the world. Its six tactical helicopter and training squadrons are spread out across the country.

- **3 Wing Bagotville** is located in Quebec's Saguenay region. It provides general purpose, multi-role, combat capable forces in support of domestic and international roles of Canada's Air Force. It also provides search and rescue missions.

- **4 Wing Cold Lake** is the busiest fighter base in Canada. It provides general purpose, multi-role, combat capable forces in support of domestic and international roles of Canada's Air Force. Home of fighter pilot training for the Canadian Forces, 4 Wing attracts Top Gun crews from all over the world to our annual air combat exercise, Maple Flag.

- **5 Wing Goose Bay** serves as a NORAD CF-18 deployed operating base and airfield supporting a mix of aviation activities, military and civilian, in North-Eastern Canada.

- **8 Wing Trenton** is the heart of Canada's air mobility forces—from delivering supplies to the high Arctic (CFS Alert) to airlifting troops and equipment worldwide. It is also responsible for search and rescue in central Canada and home to the famous Skyhawks with the Canadian Parachute Centre.

- **9 Wing Gander** is home of the 103 Search and Rescue (SAR) Squadron, providing full-time SAR services to Newfoundland and Labrador. When a call for help comes in, SAR crews at 9 Wing Gander are ready to head out in any direction from their base in Canada's most easterly province, Newfoundland.

- **12 Wing Shearwater** is the centre of naval aviation in Canada. Home of the CH-124 Sea King helicopter, 12 Wing supports the Navy with up to nine helicopter air detachments for international and domestic operations.

- **14 Wing Greenwood** is nestled in the heart of Nova Scotia's beautiful Annapolis Valley. Aurora crews conduct sovereignty and surveillance missions over the Atlantic Ocean routinely, while search and rescue capabilities are maintained 365 days of the year.

- **15 Wing Moose Jaw** is the site of the new NATO Flying Training Program in Canada (NFTC). This southern Saskatchewan town is also home to the Snowbirds, Canada's world famous aerobatic team.

- **16 Wing Borden** is the "Birthplace of the RCAF." The largest training Wing in the Canadian Forces, 16 Wing's schools offer Air Force technical training and professional development.

- **17 Wing Winnipeg** is comprised of three squadrons and six schools. It also provides support to the Central Flying School. All combined, 17 Wing turns out what are considered some of the best air navigators and multi-skilled personnel in the world. For Canadian Air Force personnel, all roads will lead to 17 Wing Winnipeg.

- **19 Wing Comox** is based on Vancouver Island. Its Aurora crews keep watch over the Pacific Ocean while its search and rescue teams regularly locate downed Aircraft in some of Canada's roughest terrain while another squadron helps train fighter pilots in tactical procedures.

- **22 Wing North Bay**, also known as the Canadian Air Defence Sector (CADS), is responsible for providing surveillance, identification, control and warning for the aerospace defence of Canada and North America at the Sector Air Operations Centre (Canada's Air Force 2006).

Each of the CAF Wings is home to one or more squadrons. A squadron is a smaller unit of personnel and equipment. In Canada's Air Force, flying squadrons are given a 400 series number (between 400–450). Each squadron has distinct purpose, aircraft, and culture. In fact, several CAF squadrons have a history stretching back to World War II.

RCAF Officers	CF Officers
Air Chief Marshal (A/C/M)	General (Gen)
Air Marshal (A/M)	Lieutenant-General (LGen)
Air Vice Marshal (A/V/M)	Major-General (MGen)
Air Commodore (A/C)	Brigadier-General (BGen)
Group Captain (G/C)	Colonel (Col)
Wing Commander (W/C)	Lieutenant-Colonel (LCol)
Squadron Leader (S/L)	Major (Maj)
Flight Lieutenant (F/L)	Captain (Capt)
Flying Officer (F/O)	Lieutenant (Lt)
Pilot Officer (P/O)	Second Lieutenant (2Lt)
Officer Cadet (O/C)	Officer Cadet (OCdt)

Other Ranks	Non-Commissioned Members
Warrant Officer Class 1 (WO1)	Chief Warrant Officer (CWO)
Warrant Officer Class 2 (WO2)	Master Warrant Officer (MWO)
Flight Sergeant (FS)	Warrant Officer (WO)
Sergeant (Sgt)	Sergeant (Sgt)
Corporal (Cpl)	Master Corporal (MCpl)
Leading Aircraftman Leading Aircraftwoman (LAC)	Corporal (Cpl)
Aircraftman First Class Aircraftwoman First Class (AC1, AW1)	Private Trained (Pte(T))
Aircraftman Second Class Aircraftwoman Second Class (AC2, AW2)	Private (Pte)

http://www.airforce.forces.gc/v2/page-eng.asp?id=634.

Figure 3.7 *Rank Structure of the Royal Canadian Air Force and the present Canadian Air Force.*

Rank Structure

Rank is an important component of military life. When a new recruit joins the forces, he or she will begin at the most junior rank and be promoted based on years of service, qualifications, performance, and courses. The better an individual performs on-the-job and in courses, the more quickly he or she will be promoted. The rank system provides a clear chain-of-command structure, through which an individual can visually identify if another person is their senior or subordinate. In this environment, junior ranks will acknowledge their respect to senior ranks by initiating a salute. The salute is an indication of the junior rank's willingness to carry out any orders delivered by the senior rank. Another important aspect of rank is salary, which is directly related to rank. Therefore, each promotion of rank also means an increase in pay.

Broadly, ranks can be segregated into two categories: commissioned officers and non-commissioned officers (NCOs). Officers are at the top of the chain of command and are typically in a management position—drawing up and giving orders. The NCO rank structure is below the officer rank structure, meaning that the highest-ranking NCO is subordinate to the most junior officer. NCOs are typically the workers who carry out the orders. A major distinction between officers and NCOs is that officers hold university degrees and NCOs do not require any formal post-secondary education. However, there are some job classifications that allow NCOs, with many years of experience, to be promoted to an officer; however this applies only to specific jobs. All pilots in the Air Force must be officers and must hold university degrees.

Air Force Careers

There are several basic requirements for joining the Air Force. All recruits must be Canadian citizens, a minimum of seventeen years old (with parental consent at that age), and meet the minimum education requirement. The minimum education requirement varies, based on the entry plan and occupation. In general, however, a Grade 10 education is required as a minimum for several NCO careers and a university degree is required for officers.

Non-Commissioned Officer (NCO) Careers

Within the Air Force, there are several important and exciting professions. As general requirements, NCO's must complete a Basic Military Qualification (BMQ) course before completing basic occupational qualification training for their career. Some of these careers include:

Meteorological Technician—observes and forecasts weather conditions

Firefighter—prevent the loss of property and life due to fire in aircraft rescue, structural, wildland, and shipboard firefighting, automobile extrication, hazardous material, and confined space/high-angle rescues

Communication Research Operator—uses a variety of sophisticated electronic equipment to intercept and analyze electronic transmissions, including foreign communications

Avionics Systems Technician—aircraft technicians who service all types of aircraft

Aerospace Telecommunications and Information Systems Technician—performs, supervises, and directs the repair and maintenance of all types of Air Force telecommunications systems and information technology infrastructure

Aerospace Control Operator—(the military version of Air Traffic Control) operates radar, computer, communications and other sensor systems for the surveillance and control of airspace. These operators must comply with Transport Canada and Federal Aviation Administration regulations. The Aerospace Control Operator training school is co-located with Nav Canada (the air traffic control agency in Canada). These operators can have very diverse careers, responsible for air traffic control (in both visual and instrument conditions) or controlling weapons systems in an air defence role at NORAD.

Officer Careers

There are several career paths in Canada's Air Force that are available only to officers. All of the following careers require a university degree. These careers include:

Aerospace Engineer: These officers are responsible for the engineering, maintenance, and management of aircraft, equipment, and facilities. These individuals are not mechanics, but exercise their technical and mechanical engineering skills to manage a very large team of highly-skilled technicians.

Air Combat Systems Officer: These officers were previously referred to as air navigators; however, due to the increasing technological and avionic sophistication of aircraft; pilots no longer require onboard navigation assistance. Therefore, the role has changed to an air combat systems officer. The systems officer has a vital role in coordinating and directing tactical missions of aircraft and crew to achieve military objectives. To accomplish their goals, these officers utilize tactical navigation systems, sensors, communication systems, electronic warfare equipment, and weapon delivery systems. A new responsibility of air combat systems officers is the management and operation of Canada's unmanned aerial vehicle (UAV) program. UAVs are power driven aircraft, other than model aircraft flown for recreational purposes, which are operated without a pilot onboard. There are several military advantages of UAVs over manned aircraft, as they can operate in risky environments, including reconnaissance over hostile territory and in bad weather conditions. Canada is currently operating Sperwer UAVs and has plans to upgrade to armed UAVs that carry air-to-ground missiles by 2012. The United States is already using armed UAVs, such as the Reaper which carries four Hellfire missiles and two 227-kilogram laser-guided bombs (McLean 2009).

CF Photo by Corporal Doug Farmers.

Figure 3.8 *The Sperwer Unmanned Aerial Vehicle (UAV), 161001, launches on its maiden flight in Kabul, Afghanistan. The Sperwer UAV is a target acquisition and surveillance drone deployed with Operation ATHENA in Afghanistan.*

Pilot: A career as a pilot within the Canadian Forces is both demanding and rewarding. If selected as a pilot, students complete primary flight training and selection in Portage la Prairie, Manitoba. Following this, they begin training on the CT-156 Harvard II in Moose Jaw, Saskatchewan and will ultimately be selected for one of three streams: helicopter (rotorcraft), multi-engine transport, or fighter.

After students have been selected into one of the three streams, their training routes diverge. Helicopter pilots will begin training on the CH-139 Jet Ranger, and multi-engine pilots will begin training on the C-90A King Air, both in Portage la Prairie, Manitoba. These courses are eight months long. If a pilot is selected to fly fighters, he or she will continue in Moose Jaw, Saskatchewan, and train on the CT-155 Hawk. This course is six to eight months long. After completion of this course, fighter pilots go to Cold Lake, Alberta, to train on the supersonic CF-18 Hornet. This training program takes approximately eight to ten months to complete. After their first training course is complete, pilots will receive their wings and will typically have the rank of Lieutenant or Captain.

CF Photo by Corporal Jean-Francois Lauzé.

Figure 3.9 *A Hawk aircraft, from 419 Squadron Cold Lake, flying over Hamilton, Ontario during the heritage flight with the B-25 Mitchell Bomber and the Avro Lancaster of the Canadian Warplane Museum of Hamilton, Ontario.*

Once pilots finish their training, they will begin work. The type of work they do depends on the stream they were selected for: rotorcraft, multi-engine transport, or fighter.

Rotorcraft pilots: These pilots may fly the:

- CH-146 Griffon light to medium utility transport helicopter for the Army

- CH-149 Cormorant Search and Rescue (SAR) aircraft for military and civilian land, air, and sea activity in Canada

- CH-148 Cyclone maritime helicopter support, used for anti-submarine and coastal defence activities, in support of Canada's Navy

Multi-engine transport: These pilots may fly the:

- CC-115 Buffalo in SAR activities

- CC-130 Hercules in tactical airlift in support of the Army or domestic or international operations

- CC-177 Globemaster allows for the rapid strategic delivery of troops and oversized combat equipment worldwide

- CP-140 Aurora and CP-140A Arcturus are maritime patrol aircraft which are used for coastal surveillance, defence, anti-submarine, drug trafficking, and territorial sovereignty patrols, in support of the Navy

- CC-150 Polaris (which is a converted Airbus A310) is used as a long-range transport aircraft that holds up to 194 passengers. One of the CC-150 Polaris aircraft is outfitted for the transportation of very important people (VIP), such as the Prime Minister of Canada, to support worldwide government operations.

- CC-144 Challenger is an executive jet, also used to transport VIP and government personnel.

Figure 3.10 *Composite image of the new CH-148 Cyclone helicopter. The Cyclone will replace the CH-124 Sea King as Canada's main ship-borne maritime helicopter.*

CF photo by Jack Szymanski.

Figure 3.11 *CC 130 Hercules during Exercise Repatriation. Exercise Repartriation practices the removal of noncombattants from a hostile area.*

Fighter: These pilots may fly the:

- CF-18 Hornet is a multi-role fighter which is used for air defence, tactical support, and training. This aircraft has had success in hundreds of military operations worldwide.

- CT-114 Tutor is an air demonstration aircraft used by Canada's famous Snowbirds.

U.S. Navy photo by Mass Communications Specialist 2nd Class (AW/NAC) Scott Taylor. (Released)

Figure 3.12a *A CF-18 from 409 Tactical Fighter Squadron flies in close formation with a US Air Force KC-135.*

CF Photo by MCpl Robert Bottrill.

Figure 3.12b *Snowbird 6 of the Canadian Forces Snowbirds, 431 Squadron Air Demostration Team, from Moose Jaw, Sask., flies over the Comox valley.*

Officer Entry Plans

To enter the military as an officer, there are two paths: Direct Entry Officer (for recruits who already have university degrees) and the Regular Officer Training Plan (ROTP). ROTP is a program that financially assists recruits with the expenses of attaining a university degree at the Royal Military College or an accredited Canadian university. If accepted into ROTP for a pilot career, the military will also pay for flight training expenses, but this is valid **only** within the Commercial Aviation Management program at the University of Western Ontario in London, Ontario. Once students have completed university, they must complete a period of obligatory service in the Canadian Forces.

ROTP Pilot Selection

ROTP pilot applicants must pass through a rigorous screening process. This screening process is incredibly important because it costs the military approximately $1.4 million to train a pilot. This is an enormous investment, so the Canadian Forces have developed an intensive selection process to find the candidates best suited for a piloting career. The selection process includes:

1. Assessments at the local recruiting office, including a medical evaluation, aptitude test, credit and criminal background check, and an interview

DID YOU KNOW?

The Commercial Aviation Management program at the University of Western Ontario is the only program where an ROTP applicant can be guaranteed a pilot occupation position in the Canadian Forces!

2. Piloting skill assessment at an air crew selection centre where applicants are tested for academic and hand-eye piloting skills on a small flight simulator. This assessment is conducted over a three-day period.

3. Last, applicants complete an intense aeromedical screening, achieving a military medical air factor of 1, at a Canadian Forces Environmental Medical establishment.

Applicants must pass each level in order to progress to the next stage. If applicants are successful, the Canadian Forces will pay subsidized education tuition, a value of approximately $20,000 over four years, and flight tuition, at a value of $65,000 over four years. Mandatory books and supplies are also included, and candidates receive a salary of approximately $70,000 over four years. The total value of ROTP is over $155,000 per student.

During the first summer, officer cadets will complete Basic Military Officer Qualifications. During subsequent summers, officer cadets will receive on-the-job training. After completing their degree, officer cadets will be sent directly to Basic Flight Training in Moose Jaw, Saskatchewan, bypassing the Primary Flight Training stage in Portage la Prairie, Manitoba.

Under the ROTP, the ratio of obligatory service is 2:1. Therefore, after completing a four-year university degree that runs eight months year, a candidate will owe the Canadian Forces sixty-four months of obligatory service. However, that service runs concurrent with the education and summer training. Therefore, after graduation only sixteen additional months are owed. For pilots, the obligatory service period will likely elapse right around the time he or she receives their wings.

Completing pilot training, whether under the ROTP or the Direct Entry Officer plan requires an obligatory service period of seven years. During the seven years of service, pilots are guaranteed relevant flying experience and an average salary of $60,000/year.

Canadian Forces Photo.

Figure 3.13

Royal Military College

The Royal Military College of Canada (RMC) was created by an Act of Parliament in 1874 to provide complete military education. RMC opened in 1876 to its first class, which consisted of eighteen cadets. Two years after it opened, Her Majesty, Queen Victoria granted the college the right to use the prefix *Royal*. In 1959, the *Royal Military College of Canada Degrees Act* was passed which allowed RMC to offer degrees in Engineering, Science, and Art (Royal Military College of Canada 2009).

There are four components of RMC life: academics, leadership, athletics, and bilingualism. It is important to note that most universities focus only on academic development through course work. RMC offers a variety of academic degree options; however the core curriculum is mixed with military knowledge. During the school year, cadets are required to wear uniforms during their academic and athletic activities. Different from a traditional university, RMC is fully residential, and officer cadets live together in a military environment following strict academic and extracurricular activity schedules. Cadets are also required to complete practical military training during summer months.

Leadership training is a requirement because, as an officer, people will be looking to the officer cadets for guidance. Therefore, all officer cadets must complete a Basic Officer Training course prior to the beginning of the academic school year. Leadership is meant to teach individuals the capacity to influence others, through instruction in leadership theory, military writing, general service knowledge, and weapons training. Leadership training is designed to develop mental and physical toughness in the cadets.

RMC supports the theory that athletic training supports the development of strong physical, mental, and leadership skills. Therefore, athletics are an important aspect of officer-cadet training. The final component of RMC life is language training. Officers must have the ability to communicate in both official languages of Canada: English and French. If cadets are not bilingual, they must attend a ten-week immersion program after completing their first year.

Canada's Air Force.

Figure 3.14a *Tomb of the Unknown Soldier*

Canadian Forces Photo.

Figure 3.14b *Poppies cover the Tomb of the Unknown Soldier on Rememberance Day.*

Military Heritage

Men and women who serve in the Armed Forces are following in the footsteps of brave soldiers who sacrificed their lives to preserve the safety and security of Canada. It is important that no one forgets the importance of this sacrifice. As a reminder of those who have given their lives in service to Canada, a Tomb of the Unknown Soldier was created in front of the National War Memorial in Ottawa. In May of 2000, the remains of an unidentified soldier who died in World War 1 was repatriated from a cemetery in the vicinity of Vimy Ridge, France. A casket carrying the Unknown Soldier's remains was then flown back to Canada, accompanied by a 45-person Canadian Forces guard, and placed to rest in a specially-designed sarcophagus in front of the War Memorial. The purpose of this Tomb is to honour more than 116,000 Canadians who gave their lives in the line of duty, and all those that die in conflict—past, present, and future (Veterans Affairs Canada 2005).

Conclusion

This chapter has reviewed the modern history of military aviation in Canada, including the unification of the three branches of the military into the Canadian Forces in the 1960s and the subsequent return to traditional values (and blue uniforms) in the 1980s. In addition, an overview of the roles of the Air Force was presented, including airspace surveillance and control, domestic and international operations, search and rescue, and humanitarian aid.

The organization of the Air Force through Wings distributed across the country was reviewed, in addition to the rank structure which defines career progression for military personnel. An overview of military occupations, including specifics on Air Force pilots, was presented. In addition, the details of the ROTP program were reviewed, which financially assists officer candidates through four years of university education who might not otherwise be able to attend university.

Serving in Canada's Air Force is much more than a job—it is a way of life. The training and discipline that are taught in the Air Force reshapes the nature of one's personality, allowing a person to succeed in rigorous and challenging environments. It is important for all Canadians to understand, respect, and pay tribute to those who contribute to the safety and security of Canada through service in the Canadian Forces.

HIGH FLIGHT

Oh! I have slipped the surly bonds of Earth
And danced the skies on laughter-silvered wings;
Sunward I've climbed, and joined the tumbling mirth
Of sun-split clouds, and done a hundred things
You have not dreamed of—wheeled and soared and swung
High in the sunlit silence. Hov'ring there,
I've chased the shouting wind along, and flung
My eager craft through footless halls of air. . . .

Up, up the long delirious, burning blue
I've topped the wind—swept heights with easy grace
Where never lark nor ever eagle flew—
And, while with silent lifting mind I've trod
The high untrespassed sanctity of space,
Put out my hand, and touched the face of God

"High Flight" is considered the sonnet of the Air Force, written by Pilot Officer John Gillespie Magee, Jr. Magee was an American who enlisted and served with the Royal Canadian Air Force. Along with hundreds of other Americans, Magee volunteered to fight, as the United States was still officially neutral. In 1941, nineteen-year-old Pilot Officer Magee was inspired to write the poem after a high altitude flight in his Spitfire. After landing, he wrote to his parents "I am enclosing a verse I wrote the other day. It started at 30,000 feet, and was finished soon after I landed." On 11 December 1941, he paid the ultimate price when his Spitfire had a midair collision over England and he fell to his death.

Pilot Officer Magee's parents, who lived in Washington, D.C., sent the poem to Archibald MacLeish, the Librarian of Congress, who acclaimed Magee as the World War II's first poet. The poem was widely reprinted and the RCAF distributed Magee's words on plaques to all training stations and airfields (Air Force Historical Studies Office n.d.).

Name _____ Date _____

PRACTICE ACTIVITY

Throughout this chapter, several of types of aircraft have been discussed. Identify the following aircraft by name and describe its primary purpose.

AIRCRAFT PHOTO	AIRCRAFT TYPE	DESCRIPTION OF OPERATIONS
	_____ _____ _____	_____ _____ _____
	_____ _____ _____	_____ _____ _____
	_____ _____ _____	_____ _____ _____
	_____ _____ _____	_____ _____ _____
	_____ _____ _____	_____ _____ _____

CHAPTER **4**

AIRPORTS
AND SECURITY

OBJECTIVES

After you have read this chapter, you will:

- Understand the importance of airports to Canadian society

- Be able to describe the organization of airports, including the National Airports System

- Appreciate the mission and initiatives of the Canadian Airports Council (CAC), including the impact of airport rent

- Understand the role of airport authorities in the management and operation of an airport

- Be familiar with the Canadian Air Transport Security Authority (CATSA) and several modern security initiatives

Introduction

Airports play a crucial role in Canadian society. Airports are much more than just a patch of pavement where aircraft land and takeoff. Airports represent a connection to provincial, national and international trade, which is crucial in a globally-integrated economy. Airports also have a powerful influence on their region. In the post World War II era, airports were pivotal in building the national transportation system in the same way that railroads did in the latter part of the nineteenth century. Currently, airports are responsible for $45 billion in economic activity in their communities (CAC 2008). For example, it has been estimated that one in twenty jobs in a city with a major airport is indirectly linked to aviation. These occupations range from transportation, hospitality (hotels and restaurants), tourism (which is a $70.6 billion industry), cargo, insurance, maintenance, government oversight, accounting, legal, and human resources, among many others. In fact, just one daily Airbus A340 flight into an airport results in the creation of approximately 660 full-time jobs (CAC 2008).

Since 1992, $9.5 billion dollars have been committed to improving the infrastructure of Canada's airports. These improvements are important on several levels. Beyond the economic importance of airports on the Canadian economy, airports are typically the first facility international guests experience when visiting Canada. In 2008, more than seven million foreign travellers passed through a Canadian airport (CAC 2008). Therefore, a country's international reputation is directly linked to the quality and upkeep of its airport.

In addition, airports allow for the transport of important cargo. In each Canadian's daily life, certain amenities are taken for granted that would not be available without the national network of airports. A few examples of these services include:

- Eating fruit or vegetables out of season
- Ordering exotic flowers in the middle of winter
- Ordering a product to be shipped overnight
- Gathering with distant relatives
- Experiencing an organ transplant or emergency medical evacuation

These examples are representative of the impact of air transportation on the lives of Canadians on a personal level. However, the impacts are more significant for Canada as a society. Air transportation allows for Canadian business people and their goods and services to be delivered internationally at unparalleled speeds. Canada, as a nation, is very trade-oriented. Canada exported $450 billion and imported $407 billion worth of goods in 2007 (CAC 2008). Without the transportation of cargo through national airports, it would be nearly impossible for Canada to sustain international trade relations.

In addition, airports also significantly impact the culture and politics of Canadian society. Airports have helped tremendously in the spread of multiple cultures and ideas, helping Canada to avoid the trappings of becoming a homogeneous nation and embrace multiculturalism. Politically, airports play a major role in bringing foreign diplomats and international business leaders to Canada with the hopes of further progress in international policy development and international trade.

All of these factors represent the positive impacts of airports on Canadian society; economically, socially, and culturally.

Airport Organization

There are 726 airports in Canada. Up until the early 1990s, most major Canadian airports were operated, owned, and subsidized by the Canadian government through Transport Canada. This placed a tremendous financial burden on the government, costing taxpayers in excess of $80 million each year (CAC 2008). In 1992, the National Airports Policy (NAP) was implemented. The purpose of the NAP was to transfer the financial and operational responsibility for airports to local authorities. To accomplish this, the NAP first categorized all Canadian airports as nationally-significant, local/regional, small, and/or remote.

Canada's NAS Airports

Calgary International Airport

Charlottetown Airport

Edmonton International Airport

Greater Fredericton Airport

Gander International Airport

Halifax-Robert L. Stanfield International Airport

Iqaluit Airport

Kelowna International Airport

London International Airport

Greater Moncton International Airport

(Montréal) Mirabel International Airport

Montreal Pierre Elliott Trudeau International Airport

Ottawa International Airport

Prince George International Airport

(Québec City) Jean Lesage International Airport

Regina International Airport

St. John's International Airport

Saint John Airport

Saskatoon John G. Diefenbaker International Airport

Greater Sudbury Airport

Thunder Bay International Airport

(Toronto) Lester B. Pearson International Airport

Vancouver International Airport

Victoria International Airport

Whitehorse International Airport

Winnipeg James Armstrong Richardson International Airport

Yellowknife International Airport

Nationally significant airports are a part of the National Airports System (NAS). NAS airports are considered to be a crucial component of Canada's air navigation system and economy as a whole. To be classified as an NAS airport, an airport was located in a capital city or had an annual traffic level of 200,000 passengers or more. There are twenty-seven NAS airports across Canada. The government legally owns all of the NAS airports; however the authority for the financial and operational management of these airports is the responsibility of the Canadian Airport Authorities (CAA). In this way, the government is viewed as the landlord for the airport, retaining ownership but limiting responsibility. The purpose of this arrangement was for the government to ensure that airports which are crucial to Canadian air transportation remain viable and operational. However, some argue that government policy has made airports less viable for several reasons discussed later in this chapter. By 2003, all of the NAS airports had been transferred from government to CAA management. This move placed all financial responsibility for airport operation and facility maintenance/growth on the CAAs.

CAAs operate under the following principles:

- CAAs are "not-for-profit" corporations, guided by a local board of directors.

- The board members are representative of the local community and do not include government employees or elected representatives.

- There are federal and provincial government representation on the board of directors.

- The method of appointment and revocation of appointments to the board of directors is specified.

- The annual general meeting is open to the public.

- All contracts in excess of a total value of $75,000 are normally awarded through a competitive bid process.

- CAAs have established community consultative committees.

- The public will has access to the CAA's key business documents.

- The CAA has a performance review conducted by an outside reviewer at least once every five years (Transport Canada 2009).

> The twenty-seven NAS airports handle 96 percent of national passenger and cargo traffic in Canada. Regional airports, although there are a total of seventy-one, handle only 4 percent.

Regional/local airports handle lower traffic volumes than NAS airports do; however, they play significant roles in local economies. The Canadian government has no ownership, financial, or operational involvement in these airports. Regional airports are owned and operated by groups in their local communities—provincial and local governments, airport commissions, private businesses, or other groups. These airports were those:

- Whose scheduled passenger traffic was less than 200,000 a year for three consecutive years

- Not the national capital or a provincial or territorial capital

- Not classified as arctic or remote airports

- With scheduled passenger traffic (Transport Canada 2009)

Additionally, there are many small airports across Canada that do not have any regularly scheduled air service. These small airports are primarily used for general aviation and recreational flying. Some of these airports are quite remote. Remote airports are those in which air transportation is the only year-round method of transportation available to the community it serves. Ownership of these airports is held by regional/local groups. However, remote airports may receive federal assistance, as their continual operation is needed to sustain the local economy.

Airports and the Environment

As an industry, aviation activities have an impact on the environment. These impacts include aircraft noise and engine emissions.

Aircraft noise impacts people who live in the area underneath the arrival and departure paths of major airports. The industry is addressing this problem, as modern aircraft are 50 percent quieter than ten years ago (www.enviro.aero 2009). In fact, it is estimated that the noise footprint of new generations of aircraft is a minimum of 15 percent lower than the last generation. In addition, ICAO introduced new noise standards for aircraft in 2006 requiring new aircraft to be one-third quieter than those built previously. ICAO estimates that between 1998 and 2004 the amount of people exposed to aircraft noise was reduced 35 percent.

Aircraft engine emissions are another major environmental concern, as it is believed that they impact global climate change. Although air travel contributes only 2 percent of worldwide greenhouse gases, the industry is committed to working towards 0 percent emissions (CAC 2008).

Canadian airports are working towards being environmentally conscious and responsible. For example, in construction projects, materials are reused and recycled. In addition, the latest heating, ventilation, and air-conditioning technologies are being used to maximize energy efficiency in airports. For example, Toronto's Pearson Airport has a co-generation system that feeds energy back into the grid. Montreal is using new lighting systems that save 80 percent of the energy used by incandescent bulbs and has installed novel heating and air conditioning units that combine electricity and natural gas and recapture emissions. Vancouver is using solar panels. The Victoria airport has implemented rules limiting idling times for taxis and is planting gardens in parking lots to improve air quality. In addition, the chemicals used at airports (such as glycol for de-icing planes) are handled responsibly to minimize environmental impacts.

The Canadian Airports Council (CAC 2008) reports that there are several myths associated with the environmental impact of aviation. Often, the public perceives aviation as being a major contributor to global pollution. However, the facts are that:

- Aviation is responsible for only 2 percent of worldwide CO_2 emissions from fossil fuel use.

- At 23 percent, transport is tied for third with agriculture for global greenhouse gas emissions, after power and land use sectors.

- Aviation is responsible for 12 percent of CO_2 emissions from all transport sources, compared to 76 percent from road transport.

- 80 percent of aviation's greenhouse gas emissions are from passenger flights exceeding 1,500 kilometres, for which there is no practical alternative.

- Today's aircraft are 70 percent more fuel-efficient than they were forty years ago, and aircraft operations have become 20 percent more fuel efficient in the past ten years.

Canadian Airports Council

Canada's Airports: Working Together, Moving Forward

Canadian airports are represented by a group called the *Canadian Airports Council* (CAC). Created in 1992, in response to the implementation of the NAP, its purpose is to be the voice of Canadian airports and lobby the Canadian government on issues of interest to airports. CAC has forty-nine members that represent 180 airports across Canada, including all NAS airports and the majority of smaller airports with scheduled passenger operations.

CAC's mission: "lead the industry through effective lobbying, timely communications and the establishment of strategic alliances with other industry stakeholders while promoting the mandate of airport authorities as the entities responsible for the efficient operation of Canada's airports. Encourage consensus building in order to effectively represent unified airport positions to government, the aviation industry and the public" (CAC 2008). CAC is led by a board of directors and managed by a team of full-time staff based in Ottawa, Ontario.

The CAC pursues several ongoing matters of importance, including airport rent, liberalized air policy ("Open Skies"), free trade zones, arrivals duty free, regulatory burden, and aviation security.

Airport Rent

The federal government considers airports to be federal assets. Therefore, the government charges local authorities rent for airport properties. Since the NAP implementation, Canadian airports have paid $2.5 billion in rent to the federal government on properties that were valued at $2 billion at the time of transfer (CAC 2008). These charges are scheduled to continue for decades. This is a very heavy financial burden for Canadian airports for which nothing is provided in return. To cover rent, NAS airports are forced to charge additional fees to airlines and passengers. Airport authorities argue that rent places Canadian airports at a disadvantage compared to U.S. airports and other forms of transportation. For example, Canadians who live just across the border from American airports (such as Buffalo, N.Y.) often choose to fly out of an American airport because the taxes and fees in Canada result in higher air fares; see Table 4.1 (CAC 2008). CAC members seek the elimination of airport rent.

TICKET COSTS AND TAXES		
FORT LAUDERDALE		
Toronto	$363	$73 taxes*
Buffalo	$245	$29 taxes*
LAS VEGAS		
Toronto	$684	$91 taxes*
Buffalo	$125	$50 taxes
*In addition to rent		

Facette, 2008.

Table 4.1 *Ticket Costs and Taxes for Toronto and Buffalo Airports*

Liberalized Air Policy: Open Skies

Currently, Canada has a liberalized air policy. This means that some statutory and regulatory constraints have been removed and economic policies do not unnecessarily limit air services. An Open Skies approach refers to air policy that goes beyond liberalization by "letting the market rule" through the elimination of aviation restrictions that are not of a safety or security nature. For example, airlines have been allowed to offer travel between international destinations and Canada since the 1930s through bilateral agreements with foreign countries. Canada has over seventy such agreements with countries across the globe (CAC 2008). However, bilateral agreements place specific restrictions on air carriers with the purpose of protecting each country's national carrier. Examples of restrictions include which airlines can operate in the country, the cities that can be served, how many times a week an airline can operate, the size of the aircraft, whether airlines can pick up passengers and carry them to a third country, and pricing rules (CAC 2008).

Open Skies eliminates these restrictions. The United States has had an Open Skies policy since the 1990s, and this approach has become increasingly popular, as Australia, New Zealand, Chile, the European Union, and Singapore have followed suit. Since 2005, Canada has negotiated several liberalized air agreements with important markets (such as China, India, Ireland, the United States, and the United Kingdom). However, Canada's international air policy remains far too restrictive for many important markets. These restrictions often result in the loss of international trade and business opportunities and place Canada at a competitive disadvantage.

If Canada were to adopt an Open Skies policy, any airline from one country can service the other. All restrictions regarding destination cities, aircraft size, or number of times an airline can operate in the country per week are lifted. Additionally, a true Open Skies policy allows an airline to pick up passengers in their home country, land to drop off and pick up more passengers in a foreign country, and continue to drop them off in a third country. True Open Skies allows for all nine freedoms of carriage.

Liberalized air agreements do **not** allow for Canada–United States cabotage. *Cabotage* refers to air service between two points in a foreign country. For example, under a liberalized air agreement, a Canadian carrier would be permitted to travel from Toronto to Las Vegas to Mexico City. However, a Canadian carrier is not permitted to travel from Toronto to Las Vegas to Los Angeles, as this constitutes cabotage. Cabotage is currently a very rare privilege; however, it is included within a true Open Skies policy.

The Canadian Airports Council supports an Open Skies policy that allows airports to remain competitive. Increasing the opportunities for international air service directly links to more trade opportunities for Canadian businesses. Therefore, an Open Skies policy has far-reaching impacts beyond passenger service to the shipment of cargo as well.

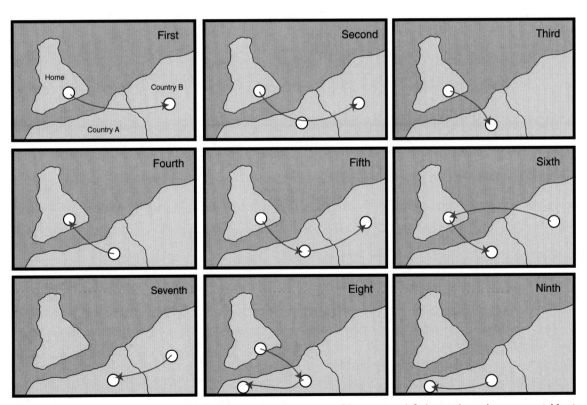

Figure 4.1 *The Nine Freedoms of the Air: This figure describes types of International flight privileges between neighboring countries.*

Free Trade Zones

To meet the needs of a globalized society, the Canadian Airports Council advocates the development of free/foreign trade zones (FTZ). FTZs are facilities that allow materials from a foreign country to be stored or processed free of duty or taxes prior to being (1) shipped to another country or (2) imported into the domestic economy. FTZs allow countries to improve competitiveness as hubs of trade and exports. These zones attract trade and transportation, and this concept is associated with economic success and growth throughout the world (CAC 2008). In fact, Canada is the only G8 country that does not have a true FTZ program. The United States has over 250 FTZs in all fifty states.

The advantages to FTZs users include improved cash flow, reduction or elimination of duties, fewer quota restrictions, avoidance of inverted tariffs, and access to economic initiatives (CAC 2008). CAC is seeking a true FTZ program for Canada.

Arrivals Duty Free

Arrivals duty free (ADF) allows international travellers to purchase products free of duty upon arrival at their destination. Approximately fifty countries around the world offer ADF. However, Canada currently does not offer ADF. This places a strain on Canadian airports, as they depend on retail operations for crucial revenue. Airport retail outlets in Canada have been struggling in recent years with the security limits on liquids and gels in carry-on luggage and the growth of ADF overseas.

Currently, Canadian federal duty-free laws (the Customs Act and Duty Free Shop Regulations) allow for the sale of duty free goods only to travellers who are preparing to leave Canada. This limitation eliminates a potential source of duty free revenue from arriving international passengers.

The Canadian Airports Council believes that the establishment of ADF would encourage passengers to purchase duty-free goods upon arrival in Canada. Given the choice, it is much more convenient for travellers to purchase duty free at the end of their trip and avoid carrying items on the aircraft. It is expected that the introduction of ADF would immediately benefit the Canadian economy, increasing sales $61 million a year and creating 400 new Canadian jobs (which results in $12 million a year in wages and $3.7 million in increased taxes) (CAC 2008).

Regulatory Burden

New environmental and security regulations place significant regulatory burdens on Canadian airports. Although security and the environment are important, each new regulation adds to the operating costs at an airport. The financial burden associated with new regulations places Canadian airports at a disadvantage compared to other airports and alternative methods of transportation. For example, aviation competes with the rail and highway for short-haul travel. Over-regulating airports makes it extremely difficult for aviation to be competitive in this market.

The CAC supports the federal Standing Committee on Transportation's 2005 interim report findings, which identified the "regulatory creep" from Transport Canada that has placed a heavy burden on airports. The findings recommended that the government cover increased costs associated with new regulations on small and regional airports.

Aviation Security

Aviation security is regulated by Transport Canada with passenger, baggage, and employee screening being the responsibility of the Canadian Air Transportation Security Authority (CATSA). CATSA is a significant partner in the "airport experience" for the travelling public. CATSA's business decisions impact the efficiency and security of the airport. However, beyond these factors, CATSA's operations directly impact airport decisions on such key issues as capital planning, retail development, and baggage and passenger flows (CAC 2008).

Funding for CATSA comes indirectly from the Air Travellers Security Charge (ATSC), which is applied to every passenger's ticket. The ATSC is administered by the Department of Finance and collected by the Canada Revenue Agency. CATSA's annual funding comes directly from government appropriations through the Consolidated Revenue Fund. Through this, the federal government allocates the amount of money CATSA receives each year. The CAC advocates that CATSA receive stable and dedicated funding that allows for the planning of future operations, equipment, and resources.

Airport Operations

Airport operations include all of the activities and personnel involved in managing the ground-based activities at an airport. Broadly, ground-based operations at major airports can be classified into one of three categories:

1. Agencies—including representatives from government agencies such as Canada Border Services, Immigration, and Transport Canada

2. Aviation and Aviation Support—including ground-side activities of airlines, ground-handling companies, de-icing services, airline fuellers, cargo-handlers, caterers, aircraft maintenance, and business and general aviation

3. Tenants and Operators—including retail concessionaires, passenger services providers, ground transportation services, and parking facilities operators

4. Airport Authority—responsible for the management and operation of the airport

Photos.com.

Figure 4.2 *The airport authority is responsible for snow removal and maintenance of ramp areas.*

At the Pearson Toronto International Airport, there are approximately 30,000 ground-side workers across the various operational sectors. With so many employees and customers, airports can be thought of as self-contained cities. Understanding the intricacies of airport operations is beyond the scope of this text. However, an overview of various operational aspects specific to airport authorities will be presented.

The airport authority is responsible for most of the activities required for the safe, secure, and efficient operation of the airport. Some examples of these activities include:

• Customer service

• Management and maintenance of facilities

• Operation of terminals

- Provision of services inside terminals
  - Commercial and retail services to the public
  - Flight information displays
  - Maintenance of common use areas
- Operation of parking garages and ground-side systems
- Operation of airside (through interface with NAV CANADA)
- Maintenance of airside (snow clearing and surface maintenance)
- Support activities
  - Finance
  - Planning
  - Construction
  - Project management
  - Marketing
  - Legal
  - Public affairs
  - Purchasing
  - Human Resources

Historically, individual airports relied upon the support services provided by the regional and headquarters offices of Transport Canada. These included finance, legal, public affairs, and purchasing, among many others. Therefore, the management group at the airport was very operationally focused. The management would focus almost exclusively on the core activities of airport operations, facility maintenance, safety, and security. Since the implementation of NAP, the support services from Transport Canada are no longer available. In response, airports have evolved into self-contained businesses which operate several support departments. In addition to core activities, airports now have a range of support functions, including legal, public affairs, purchasing, construction and development, strategic planning, marketing, air services marketing, corporate affairs (including corporate social responsibility), and customer experience. Modern airports are continually tracking and modifying operational strategies, measuring customer experiences, and developing new products and services. As it faces global competition, the airport business is more fluid, competitive, volatile, and dynamic than it has been in the past.

Airport Costs

Airports are continually working to increase revenue sources to cover their costs. These costs include:

- Rent: All NAS airports are required to pay rent to the federal government. In 2009, the Greater Toronto Airport Authority is required to pay $145 million in rent. This places a tremendous burden on the Canadian airports.

- Payments in lieu of property taxes to the municipality

- Debt service costs: As self-financing and self-sustaining operations, airports borrow money through operating lines of credit and issue bonds.

Airport Revenue Generation

Although they are not-for-profit corporations, airport authorities at NAS airports must be continually focused on cost-efficiency to cover their costs and ensure that they remain competitive with other airports in their region. NAS airports aim to operate on a break-even basis after covering the ground rent, payments in lieu of property taxes, and debt service costs. Airport authorities have two main sources of revenue generation. These two streams are aeronautical and non-aeronautical.

Aeronautical Revenue

Aeronautical revenue includes landing fees and general terminal fees.

Landing fees are levied against airlines based on the maximum takeoff weight of each aircraft that lands at the airport. Landing fees are used to cover the projected operating costs associated with the airfield and ground-side areas of the airport.

> *Landing Fees*—based on maximum takeoff weight of aircraft that land at the airport
>
> *Terminal Fees*—based on seating capacity of aircraft that dock at the terminal

Terminal fees are meant to recover the operational costs of maintaining common areas of the terminal (including gate areas, lounges, and restrooms). Airlines are assessed terminal fees based on the seating capacity of each aircraft that docks at the terminal. The terminal fee is not based on the number of occupied seats, but the total number of seats on the aircraft, whether empty or occupied.

Photos.com.

Figure 4.3 *This aircraft will be charged a terminal fee for docking at the airport.*

Not all types of aircraft are the same. For example, all Airbus 330s are not identical in weight and seating capacity. Airlines typically choose the equipment and seating configuration from the aircraft manufacturer when they purchase a new aircraft. Therefore, airport authorities reference the aircraft's registration to determine the number of seats and maximum takeoff weight in the calculation of fees.

Non-aeronautical Revenue

Non-aeronautical revenue includes all sources of revenue at the airport besides landing and terminal fees. There are many sources of non-aeronautical revenue, including:

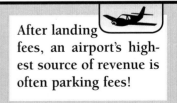

> After landing fees, an airport's highest source of revenue is often parking fees!

- Concessions and retail activities (such as duty-free shops and bookstores)

- All types of advertising

- Parking (which is the highest source of revenue after landing fees)

- Ground transportation fees

- Land rentals for hangars and other facilities

- Airport improvement fees (which may be applied to every passenger ticket except connecting passengers)

Photos.com.

Figure 4.4 *These icons demonstrate the range of services that are available within an airport, many providing non-aeronautical revenue.*

Airports are continually striving to increase the ratio of non-aeronautical revenue to aeronautical revenue. The primary challenge in airport management is to drive revenues and reduce costs. To drive revenues, airports cannot continually raise landing and terminal fees to airlines. Maintaining the lowest-possible landing and terminal fees is important to allow airports to remain competitive.

The reason for this is that airports are no longer monopolies. Airport fees must be competitive with other nearby airports. For example, Toronto is in direct competition with Buffalo, N.Y., Hamilton, ON, and London, ON. On a larger scale, airports compete for gateway traffic. Gateway traffic is intercontinental traffic that uses a Canadian airport as a gateway to connect with another destination. Examples include Europe to North America, Europe to South America, Asia to South America, and Europe to Asia. Gateway traffic is long-haul, premium, higher-yield traffic over which airports compete. In some airports, the majority of traffic is connecting, such as Chicago, Ill. However, many Canadian airports, including Toronto International, have very strong origin-destination traffic which serves the local economy. At Toronto International 75 percent of traffic is origin destination and 25 percent is connecting.

Cost Approaches

There are three cost approaches that are used by airports: residual, compensatory and a hybrid of these two approaches.

Residual Methodology

Within this approach, airlines pay any airport operating costs that are not covered by non-aeronautical revenue. For example, at the Toronto Pearson International Airport, a residual methodology is in place. Landing fees and terminal fees are offset by projected non-aeronautical revenue. Therefore, if a financial surplus has been produced, this money is returned to the airlines through reductions in landing and terminal fees. In a residual approach, the airlines assume a level of risk, because, if non-aeronautical revenue is insufficient to cover costs, their fees will increase. However, the benefit to airlines is that any surpluses created through non-aeronautical revenue are given back to the airlines. For the airport, this approach can make development projects difficult because all extra revenue is given back to airlines.

Compensatory Methodology

Within this approach, airports charge airlines fees equivalent to the amount necessary to recover operating costs. For example, the fees charged to an airline are based specifically on the facility or service utilized rather than a general airport-wide cost. In this manner, the airport assumes the risk of a potential revenue shortfall. Therefore, from an airport perspective, the compensatory approach lacks the security of a residual approach. However, any surplus profits are retained by the airport.

Hybrid Methodology

This approach is also known as a cost centre approach. It represents the middle ground between residual and compensatory methodologies. In general, a hybrid approach will exclude some non-aeronautical revenue from the residual cost pool. So the airfield may remain in the residual pool while some non-aeronautical revenue sources do not. A hybrid approach provides airports with more control over its funds than a residual methodology but less control than a compensatory approach. A hybrid approach is increasingly common in modern airports, as more than half of major airports in the United States and Canada utilize this approach (ACI 2003).

Customs

A rather unique challenge faced by Canadian airports is preclearance of United States (U.S.) customs and immigration. Preclearance allows passengers travelling to the U.S. to clear customs, immigration, and agriculture prior to leaving their departure airport. In this system, passengers are technically on U.S. territory when they reach their departure gate to board their aircraft. This system offers a tremendous benefit to U.S. carriers travelling back to their hubs because there is no requirement for customs clearance upon arrival. This is also a benefit to Canadian air carriers, as it allows for the development of point-to-point markets with smaller aircraft that travel to small U.S. airports with no customs services.

The airport authority provides the facilities for U.S. customs, and costs are covered by air carriers. The U.S. does not pay for this service—in fact, the Canadian airport pays for the privilege!

However, offering preclearance within an airport results in added costs and complexity. The reason for this is that the terminal must be designed to separate three sectors of traffic: domestic, international, and preclearance. Other major airports, in the U.S. or in Europe, have only two sectors: domestic and international. Designing a terminal that segregates the three sectors of traffic within a single building without having passenger flows become too complicated is tremendously challenging. In fact, the necessity of separating the three sectors of traffic is one reason why airports have long walking distances.

One sector of traffic that airports are trying to drive is international connections, including both in-transit preclearance and international-to-international passengers. The former are arriving from international destinations and are connecting through a Canadian airport to a destination in the U.S., while the latter are connecting from an international point to other international locations. In-transit preclearance and international-to-international passengers make airport planning and operation more challenging and costly. Multilevel facilities are required because the Canadian government will not allow the co-mingling of arriving and departing passengers.

Job Title:

Airport Operations

Career Paths:

Careers within airport management often begin with airport internships for students completing their college or university education. After graduation, there are many entry-level positions available within airport operations, including operations coordinator, dispatcher, terminal operations specialist, and facility allocation resource planner. From there, individuals will advance to management positions within the organization.

Salary:

- Entry-level—starting at $35,000
- Management—starting at $55,000
- Employees are typically unionized, and wage levels are governed by collective agreements.

Training:

Many positions are filled by graduates from college- or university-level aviation programs.

Lifestyle:

As the types of positions within this sector are numerous, the lifestyle associated with each position varies. Entry-level positions, such as at the airport operations centre, may work night shifts, as it operates twenty-four hours every day. A terminal operations specialist will monitor operations on the airport floor and provide assistance to the public. A resource planner will work with airline slot coordinators to allocate runway slots and terminal building gate allocations. From these positions, individuals will work their way up to management roles within various sectors of airport operations.

Security

In Canada, aviation passenger, baggage, and employee security is the responsibility of the Canadian Air Transport Security Authority (CATSA). This section will explore the background, organization, and responsibilities of CATSA along with various technologies that are used to enhance airport security. It is important to keep in mind that the challenge facing aviation security professionals is to balance the often conflicting requirements of passenger privacy, efficient screening, and cost-effective measures against the ultimate goal of security (ICAO 2005). For example, one could argue that the highest level of security would be achieved by personally interviewing and strip-searching every airline passenger. However, passengers would be in an uproar over the invasive techniques, and the process would be **so** slow and labour intensive that only a handful of passengers would be cleared each day. Clearly, this would not be an appropriate aviation security strategy! Instead, security professionals are continually investigating ways to maximize security while maintaining privacy, efficiency, and cost.

Canadian Air Transport Security Authority (CATSA)

The tragic events of 11 September 2001 impacted all sectors of the aviation industry. However, the greatest impact was probably felt in the security sector of aviation. In response to this terrorist incident, the government of Canada invested $2.2 billion in aviation security initiatives, which included the development of the Canadian Air Transport Security Authority (CATSA). CATSA came into force on 1 April 2002, through Bill C-49. CATSA reports to Parliament through the Minister of Transport Infrastructure and Command. Currently, CATSA is partially funded indirectly through security charges that are placed on every ticket. These charges range from $5 for intra-provincial travel to $17 for international trips.

CATSA is a Crown corporation and is based in the Ottawa, Ontario, region. Four training centres are operated by CATSA to teach staff how to conduct pre-board screenings. CATSA training centres are located in St. John's, Newfoundland; Halifax, Nova Scotia; Winnipeg, Manitoba; and Edmonton, Alberta. However, training takes place in other cities as well. The mandate of the organization is the delivery of consistent, effective and highly professional service that is set at or above the standards established by Transport Canada. The organizational structure of CATSA includes an eleven-member board of directors, including a chairperson who is appointed by the Governor in Council based on a recommendation from the Transport Minister. Two board members are proposed by airport operators, and another two are proposed by representatives from the airline industry. The purpose of the board is to provide strategic direction, financial oversight, corporate oversight, and good governance. In addition to the board of directors, CATSA has strong executive team, including a president/CEO, two senior vice-presidents (people and operations), and three vice-presidents (strategic and public affairs, chief technology officer, and chief financial officer).

CATSA's vision is to be a world leader in air transportation security through operational and corporate excellence.

CATSA's responsibilities fall into four major areas:

1. Pre-board screening of passengers and their belongings. (However, CATSA contracts screeners from private security companies.)

2. Hold baggage screening, which includes the acquisition, deployment, operation, and maintenance of explosives detection systems at airports

3. Implementation of a restricted area identification card

4. Screening of non-passengers entering airport restricted areas

Pre-board Screening

When one thinks of airport security, pre-board screening (PBS) is what comes to mind. PBS involves screening all passengers and their belongings for prohibited items. On 31 December 2002, CATSA assumed full operational responsibility for PBS in Canadian airports.

PBS typically involves x-ray scanning of carry-on luggage, walk-through metal detectors, explosive trace detection technology of baggage, and physical searches of persons of interest. The photo shows a typical airport security screening process, including x-ray machines for carry-on baggage and walk-through metal detectors.

Figure 4.5 *Passengers walking through pre-board screening (PBS).*

SHOE BOMBER

On 22 December 2001, Richard Reid was aboard American Airlines flight 63, a Boeing 767 travelling from Paris to Miami. Midway through the flight, passengers alerted the flight attendants that they smelled smoke. While searching for the source of the smell, a flight attendant identified the smell as coming from Richard Reid. After approaching him to remind him of the non-smoking rule onboard the aircraft, she noticed that he had a shoe in his lap and that he was trying to light a fuse extending from the shoe. Richard Reid fought with the cabin crew, but was eventually restrained by other passengers onboard who stepped in to help. A doctor who was a passenger gave Reid a sedative from the airplane's medical kit. Flight 63 diverted to Boston, escorted by two fighter jets, without further incident.

Investigation revealed that Reid's shoe bomb was actually quite sophisticated. The shoe contained two types of explosives and was powerful enough to blow a hole in the airplane's fuselage. Based on a palm print and hair found in the shoe's explosives, investigators concluded that Reid had help constructing the shoe-bomb. Further investigation revealed his linkages with al-Qaeda, a terrorist organization. For his crime, Reid was sentenced to life in prison.

The ramifications of this event are still being felt in the world of aviation security. As a result of this incident, passengers are now asked to remove their shoes while passing through PBS and place them on the x-ray scanner. Although it's common to hear passenger complaints during this process, which is perceived as an inconvenience, it is a small price to pay for enhanced security.

Backscatter X-ray

To enhance the effectiveness of PBS, a new technology called *backscatter x-ray* has been implemented in several airports across the United States, and pilot-tested in Kelowna, B.C., to provide security screeners with detailed pictures through passengers' clothing with the goal of detecting concealed weapons. This system is meant to allow screeners to identify non-metallic objects, including explosives, which are not detected by traditional metal detectors. This scan is meant to replace hands-on pat downs for people after they have set off the metal detector or if they are identified as a person of interest.

This system is controversial, often referred to in the media as a *virtual strip search*. Many individuals and privacy groups consider this technique to be an invasion of privacy that should be used only as a last resort. However, advocates of this technology argue that it will speed up the efficiency and effectiveness of PBS (Wong 2008). Figure 4.6 is an example of what a backscatter x-ray image may look like. However, in this figure the man's undergarments are seen. In actual backscatter images, the outline of his genitals would be visible. Some backscatter systems are able to blur private areas in the image, although some argue that this decreases the effectiveness of the system, as people could hide prohibited items in these areas.

> **Ask yourself:**
>
> **Which is more important to you, your security or your privacy?**

To improve privacy, the scanners blur the faces of passengers. However, other sensitive areas of the body (such as breasts or genitals) are not blurred because it may limit the screener's ability to identify concealed weapons. The security screener viewing the images is remotely located where he or she cannot see the passengers being scanned. There is no ability to print images, although they can be retained until no longer pertinent. More recently, the images produced through backscatter x-ray have been modified to resemble chalk outlines rather than nude pictures. Refer to the photos of backscatter images that present the modified chalk outline pictures.

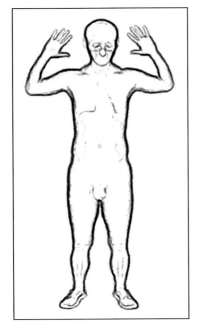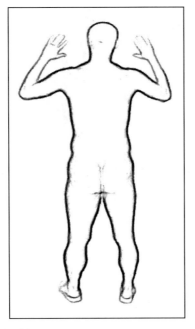

http://www.tsa.gov/research/privacy/backscatter.shtm.

Figure 4.6 *An example of a male "chalk outline" backscatter x-ray picture.*

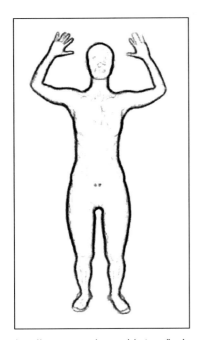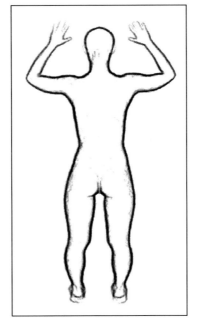

http://www.tsa.gov/research/privacy/backscatter.shtm.

Figure 4.7 *An example of a female "chalk outline" backscatter x-ray picture.*

These are actual images that would be shown to a security scanner. Although this technology is controversial and sparks a great deal of debate, it is likely that the usage of backscatter x-rays will increase in the future as passengers demand more expeditious and efficient screening processes.

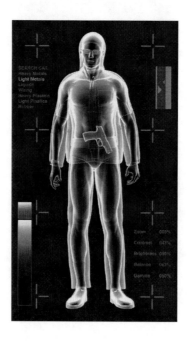

Figure 4.8 *An example of a "virtual strip search."*

Hold Baggage Screening

Another CATSA responsibility is hold baggage screening (HBS). ICAO Amendment 10 to Annex 17 mandated that 100 percent of bags loaded onto originating international flights must be screened as of 1 January 2006. This security measure was introduced to protect the travelling public from potential explosives/dangerous goods in checked baggage. While a passenger is experiencing PBS, their checked baggage, which they placed on a conveyor belt earlier in the security process, is experiencing HBS. The major challenge associated with this is to screen baggage effectively, as early x-ray technologies were associated with significant shortfalls in technology and the ability of human operators to detect dangerous goods. Currently, there are several options available for HBS. These options, as described by ICAO (2005), include:

Physical Search: This involves security personnel inspecting each piece of checked luggage by hand. This option is slow, invasive to passenger privacy, and labour intensive. In the search for concealed bombs, screeners may have to destroy personal property to find the item, or it may escape detection altogether. This approach is best used when combined with information from other approaches.

Canine: Explosive detection dogs, often called K-9s, were security professionals' first means of identifying explosive devices. The effectiveness of the dog is dependent on the level of training and readiness of the dog and its handler, the frequency of recurrent training, and the care given to handling and maintaining training aids. For example, training aids may become contaminated with other scents, including the handler themselves, rubber gloves, or the storage container. This option carries high maintenance and training costs, and dogs can work only for short periods at a time. Some experts argue that canine searches are better suited for airport/aircraft searches rather than routine HBS.

Trace Detection: Canada, in partnership with the United Kingdom and the United States, conducted tests in the late 1990s to identify the expected contamination levels associated with terrorist bombs in hold bags. Based on the results of these tests, a "mixed" screening strategy was developed. This requires that a number of bags be tested in the following way: exterior only, exterior and interior, and complete screening of any objects large enough to conceal a bomb. This type of screening carries high labour costs. In addition, processing times are long because each bag has to be handled by a screener. If the sampling process is implemented incorrectly, dangerous items will be missed. Currently, only smaller airports use the trace-based method.

Conventional X-ray: This technology has been greatly improved since it was introduced in the 1970s. Modern x-rays provide the screener with enhanced information about baggage contents. Organic materials (such as explosives and the vast majority of baggage contents) are presented in one colour, and metallic objects are presented in a second colour. However, the task of visually identifying a bomb is very challenging, as they range greatly in shape. The results of surprise testing are that the system does not always attain a high level of detection (ICAO 2005). The effectiveness of this method is completely dependent on the motivation, training, and vigilance of the screener.

Advanced Technology: Advanced technology (AT) x-ray machines process bags more quickly and are capable of automatic detection of explosive items. AT systems often utilize multiple transmission views and dual energy. However, the United States has repeatedly determined that AT systems do not always meet detection standards established by ICAO (ICAO 2005). Therefore, AT systems create a security gap which could be exploited by terrorists.

Explosive Detection Systems: Computed tomography based explosive detection systems (EDS) are the best method of screening hold baggage. These systems are large, heavy, and expensive (over 1 million dollars Canadian) and must be integrated into the baggage handling system at the airport. However, they provide a high level of detection with a manageable false alarm rate. Overall, EDS provide the highest level of security to the travelling public.

CATSA has the responsibility for purchasing, integrating, operating, and maintaining EDS equipment at specific airports across the country. The airports where CATSA has implemented EDS handle 99 percent of Canadian air traffic. The government of Canada continues to commit funds to allow for further integration of EDS into airports.

Restricted Area Identity Card

An enhanced restricted area identification card (RAIC) has been implemented by CATSA. The RAIC is used by non-passengers, such as ground crew or baggage handlers, at the airport and integrates biometric identifiers (such as fingerprint or retinal scanning). The purpose of the RAIC is to enhance the security of restricted airport areas at all major Canadian airports. The validity of each card will be enhanced through the use of a national database to allow for the authentication of RAICs. Currently, over 100,000 RAICs have been issued to staff in airport terminals. This adds an extra layer of security to restricted airport zones.

Non-Passenger Screening (NPS)

Transport Canada regulated non-passenger screening (NPS) in 2004 to add a layer of scrutiny to Canada's air transport security system. NPS screens people and goods seeking access to restricted areas in an airport. Non-passengers include flight crews and all airport workers (caterers, maintenance, and baggage handlers). NPS enhances airport and civil aviation security through random and unpredictable security checkpoints at entry points to or within airport restricted areas at Canada's twenty-nine largest airports. Over one thousand random NPS of people and goods occur each day at airports across Canada.

Airport Security

Airport authorities also play an important role in airport security. Airport authorities typically have their own security group, which manages the issuance of restricted area passes. In addition, airports operate and maintain a large network of closed-circuit televisions, all forms of access control (including doors, signage, and fencing), and are responsible for maintaining the integrity of primary security lines.

The airport provides the space for CATSA passenger pre-screening activities and hold baggage screening systems, although CATSA is responsible for passenger security. Airport authorities construct security screening facilities on a cost-recovery basis from CATSA. In addition, it is important to note that airport authorities work directly with CATSA to maintain a free-flowing security system. This is crucial, because any delays in airport security impact the entire airport system and the customer experience.

Conclusion

Airports represent a connection between Canada and the rest of the world. This linkage is responsible for trade opportunities, political partnerships, immigration, and tremendous economic stimulus to the Canadian economy. However, as the Canadian Airports Council identifies, there are several issues that threaten the ability of Canadian airports to remain competitive in a global market. The most significant of these issues is the annual cost of airport rent. Airport authorities are continually striving to overcome this burden by driving non-aeronautical and aeronautical revenue sources. This allows airports to reduce landing and terminal fees levied on airlines and remain competitive for important gateway traffic. In addition, CATSA is working within airports to ensure the security of the system. Security measures include implementing modern pre-board and hold baggage screening technology. Overall, Canadian airports are striving to provide secure, efficient, and enjoyable experiences for passengers through partnerships with airlines, ground-operations companies, government agencies, and service providers.

PRACTICE ACTIVITY

Ask yourself how security innovations impact privacy, efficiency, or cost, and how these are weighted against the ultimate goal of security. Put yourself in the role of an airport professional. Describe how important each factor is to you personally. Next, write down whether you would be willing to sacrifice each factor in the name of improved security. For example, would you sacrifice your privacy through invasive pat-downs or backscatter x-rays, endure long lines requiring you to arrive at the airport an hour earlier, or pay higher ticket prices to improve the effectiveness of airport security?

	DESCRIBE HOW IMPORTANT EACH FACTOR IS TO YOU.	WOULD YOU BE WILLING TO SACRIFICE THIS FOR THE SAKE OF IMPROVED SECURITY? DESCRIBE YOUR REASONING.
Privacy		
Efficiency		
Cost-effectiveness		

Describe or create a system that reflects the middle ground between privacy, efficiency, and cost-effectiveness and security.

CHAPTER 5

AIR LAW IN CANADA

OBJECTIVES

After completing this chapter, you will:

- Understand the challenges and necessity of regulating aviation on an international scale
- Be able to describe the importance of the Chicago Convention
- Know the history, structure, and purpose of the International Civil Aviation Organization (ICAO)
- Understand who has jurisdiction over aviation in Canada
- Be familiar with Transport Canada and the *Canadian Aviation Regulations* (CARs)
- Know the purpose of the Transportation Appeal Tribunal in Canada
- Understand the process of the Canadian Aviation Regulation Advisory Council (CARAC)

International Aviation Regulations

Aviation is an industry that requires international cooperation. International aviation could not survive if every country regulated air travel in a unique and independent manner. For example, passengers and airlines need to trust that baggage security screening will be conducted with the same level of scrutiny in Canada as in every other part of the world. In order to ensure standardized aviation regulations around the globe, countries must agree to enforce the same regulatory principles. This can be a very difficult task.

Chicago Convention

The foundation for aviation regulations on an International scale was established in 1944. Due to World War II, the aviation industry had advanced at a rapid rate. "At that time, a vast network of passenger and freight carriage was set up, but in order for air transport to support and benefit at peace, there were many obstacles, both political and technical, to overcome. In the early days of 1944, the Government of the United States conducted exploratory discussions with other allied nations to develop an effective strategy. On the basis of these talks, 52 States met in Chicago in November 1944" (ICAO n.d. 2). Over the course of the next five weeks, representatives from the States discussed the many issues of international aviation. The result of their work was the Convention on International Civil Aviation, which is also called the *Chicago Convention*. Although the convention was held in 1944, it did not come into effect until twenty-six countries had ratified it in April of 1947.

There are ninety-six articles contained in the Convention that describe the privileges and restrictions of States. In addition, details on the provision and adoption of International Standards and Recommended Practices (SARPs) were included. The Convention established that every State has sovereignty over the airspace above their territory and that no airline may operate in that airspace without consent from the State (ICAO n.d.).

Currently, of the 195 countries in the world, 190 countries are a part of the Convention. This is important because, if a country is not part of the Convention, there won't be any international air travel to or from the country.

The Convention consists of four parts:

1. Air Navigation—covering general principles and application of the Convention, flight over territory of contracting States, nationality of aircraft, measures to facilitate air navigation, conditions to be fulfilled with respect to aircraft, and international standards and recommended practices

2. International Civil Aviation Organization (ICAO)—covering the organization, assembly, council, air navigation commission, personnel, finance, and other international arrangements

3. International Air Transport—covering information and reports, airports and other air navigation facilities, and joint operating organizations and pooled services

4. Final Provisions—covering other aeronautical agreements and arrangements, disputes and default, war, annexes, ratifications, adherences, amendments, and denunciations, and definitions

The annexes are an important component of the convention. Annexes are added to the Convention when something needs to be updated. For example, environmental protection was added to the convention in annex 16. This is an issue which is extremely important in modern aviation but was not considered in the 1940s. The reason for adding annexes, rather than creating a new convention, is that it is extremely difficult to get countries to agree and sign on to a convention. So the annexes are used to modify or update the existing Convention. In this manner, States that have already signed on to the Convention must abide by the annexes. Currently, there are eighteen annexes included in the Convention:

1. Personnel Licensing

2. Rules of the Air

3. Meterological Service for International Air Navigation

4. Aeronautical Charts

5. Units of Measurement to be Used in Air and Ground Operations

6. Operation of Aircraft

7. Aircraft Nationality and Registration Marks

8. Airworthiness of Aircraft

9. Facilitation

10. Aeronautical Telecommunications

11. Air Traffic Services

12. Search and Rescue

13. Aircraft Accident and Incident Investigation

14. Aerodromes

15. Aeronautical Information Services

16. Environmental Protection

17. Security—Safeguarding International Civil Aviation Against Acts of Unlawful Interference

18. The Safe Transport of Dangerous Goods by Air (ICAO n.d. 8)

> **"WHEREAS THE FUTURE DEVELOPMENT OF INTERNATIONAL CIVIL AVIATION CAN GREATLY HELP TO CREATE AND PRESERVE FRIENDSHIP AND UNDERSTANDING AMONG THE NATIONS AND PEOPLES OF THE WORLD, YET ITS ABUSE CAN BECOME A THREAT TO THE GENERAL SECURITY; AND**
>
> > **"WHEREAS IT IS DESIRABLE TO AVOID FRICTION AND TO PROMOTE THAT COOPERATION BETWEEN NATIONS AND PEOPLES UPON WHICH THE PEACE OF THE WORLD DEPENDS;**
> >
> > **"THEREFORE, THE UNDERSIGNED GOVERNMENTS HAVING AGREED ON CERTAIN PRINCIPLES AND ARRANGEMENTS IN ORDER THAT INTERNATIONAL CIVIL AVIATION MAY BE DEVELOPED IN A SAFE AND ORDERLY MANNER AND THAT INTERNATIONAL AIR TRANSPORT SERVICES MAY BE ESTABLISHED ON THE BASIS OF EQUALITY OF OPPORTUNITY AND OPERATED SOUNDLY AND ECONOMICALLY;**
> >
> > **"HAVE ACCORDINGLY CONCLUDED THIS CONVENTION TO THAT END."**
>
> *PREAMBLE TO THE CONVENTION ON INTERNATIONAL CIVIL AVIATION*
>
> *SIGNED AT CHICAGO, ON 7 DECEMBER 1944*

International Civil Aviation Organization

With the creation of the Chicago Convention, the International Civil Aviation Organization (ICAO) was formed. ICAO is a United Nation's special organization that has been headquartered in Montreal, Quebec, since 1945. ICAO is the global forum for civil aviation and works to achieve safe, secure, and sustainable aviation through international cooperation. ICAO's strategic objectives include safety, security, environmental protection, efficiency, continuity, and rule of law (ICAO 2004).

The objectives of ICAO are to develop: "the principles and techniques of international air navigation and to foster the planning and development of international air transport so as to:

(A) INSURE THE SAFE AND ORDERLY GROWTH OF INTERNATIONAL CIVIL AVIATION THROUGHOUT THE WORLD;

(B) ENCOURAGE THE ARTS OF AIRCRAFT DESIGN AND OPERATION FOR PEACEFUL PURPOSES;

(C) ENCOURAGE THE DEVELOPMENT OF AIRWAYS, AIRPORTS, AND AIR NAVIGATION FACILITIES FOR INTERNATIONAL CIVIL AVIATION;

(D) MEET THE NEEDS OF THE PEOPLES OF THE WORLD FOR SAFE, REGULAR, EFFICIENT AND ECONOMICAL AIR TRANSPORT;

(E) PREVENT ECONOMIC WASTE CAUSED BY UNREASONABLE COMPETITION;

(F) INSURE THAT THE RIGHTS OF CONTRACTING STATES ARE FULLY RESPECTED AND THAT EVERY CONTRACTING STATE HAS A FAIR OPPORTUNITY TO OPERATE INTERNATIONAL AIRLINES;

(G) AVOID DISCRIMINATION BETWEEN CONTRACTING STATES;

(H) PROMOTE SAFETY OF FLIGHT IN INTERNATIONAL AIR NAVIGATION;

(I) PROMOTE GENERALLY THE DEVELOPMENT OF ALL ASPECTS OF INTERNATIONAL CIVIL AERONAUTICS" (ICAO 2006 20).

ICAO Organization

ICAO is made up of a sovereign body called the *Assembly* and a governing body called the *Council*.

The Assembly

All countries which have signed onto the Convention have an equal right to be represented at the meetings of the Assembly, and each is entitled to one vote. The Assembly meets at least once every three years, convened by the Council, at a suitable time and place. Decisions of the Assembly are taken by a majority of the votes cast. During meetings of the Assembly, several actions will be taken, including election of the president and officers of ICAO as well as the states represented on the Council, reports of Council will be examined, annual budgets will be voted upon, and expenditures will be reviewed, in addition to other activities (ICAO 2006).

ICAO. Retrieved from http://www.icao.int/icao/photos/.

Figure 5.1 *The First Meeting of the ICAO Assembly held at the Windsor Hotel, Montréal, Quebec from 21 May to 07 June 1946.*

ICAO. Retrieved from http://www.icao.int/icao/en/assembl/a36/photos/.

Figure 5.2 *36th Session of the ICAO Assembly, 18 September 2007.*

The Council

The Council of ICAO is a permanent body responsible to the Assembly. It is composed of thirty-six States elected by the Assembly for a three-year term. Canada is a member of the ICAO Council. Priority regarding which countries are chosen to be members of the Council are given to (1) the States of chief importance, (2) States that contribute the largest to international civil air navigation facilities, and (3) States that might not otherwise be included but that are located in geographic areas not otherwise represented on the Council. No Council representatives may be actively associated with or financially interested in an international air service operation. The Council is lead by a president who is elected for a three-year term.

ICAO. Retrieved from http://www.icao.int/photos

Figure 5.3 *A Meeting of the First Council of ICAO in Montréal, 1946. Canada is represented on the Council, 2nd delegate from the right.*

There are numerous activities the Council pursues, including submitting annual reports to the Assembly, carrying out the directions of the Assembly, determining rules of procedure, appointing an Air Transport Committee, establishing an Air Navigation Commission, administering finances, determining the emoluments of the President of the Council, appointing a chief executive officer (who is called the Secretary General), collecting and publishing information relating to the advancement of air navigation and operations, reporting infractions of the Convention, and designating Annexes to the Convention (ICAO 2006). Perhaps the most important duty of the Council is to adopt International Standards and Recommended Practices and to incorporate them, as annexes, into the Convention.

MEMBERS OF THE ICAO COUNCIL

Argentina

Australia

Austria

Brazil

Cameroon

Canada

Chile

China

Colombia

Egypt

Ethiopia

Finland

France

Germany

Ghana

Honduras

Hungary

India

Italy

Japan

Lebanon

Mexico

Mozambique

Nigeria

Pakistan

Peru

Republic of Korea

Russian Federation

Saint Lucia

Saudi Arabia

Singapore

South Africa

Spain

Tunisia

United Kingdom

United States

Canadian Aviation Regulations

Transport Canada

ICAO works diplomatically to set international standards for aviation. However, countries have the final say and can choose to opt out of ICAO standards and recommended practices (SARPs). Therefore, each country must designate a group which has domestic authority over aviation regulations. Transport Canada (TC), and ultimately the Minister of Transport, is responsible for the policies and programs surrounding air, marine, road, and rail transportation in Canada. TC works to ensure the maximum level of safety, security, efficiency, and environmental responsibility in all modes of transportation, including civil aviation.

CASE

INTERNATIONAL AVIATION REGULATIONS—WITHOUT CONSISTENCY, COMPLICATIONS ARISE

The COSPAS–SARSAT system is an international satellite system for search and rescue. In aviation, search and rescue (SAR) workers rely on a signal from an aircraft's emergency locator transmitter (ELT). An ELT is required aircraft equipment that automatically transmits an emergency signal in the event of a crash. Originally, ELTs operated on a 121.5 MHz frequency. This was established in the 1950s as the international aircraft emergency frequency. However, there are several shortcomings of 121.5 MHz ELTs; this resulted in the requirement of all aircraft in Canada to upgrade their ELTs to 406 MHz models. This decision was made in 2000 in response to guidance from ICAO. As of 1 February 2009, the COSPAS–SARSAT system no longer monitors ELTs on the old 121.5 MHz frequency.

406 MHz ELTs provide a digital signal and have many advantages over the old 121.5 MHz analog signal. For example:

- The 406 MHz ELT includes a unique identification function which allows search and rescue to know the make and model of the aircraft before arriving at the accident site.

- The 406 MHz ELT provides global coverage, whereas the 121.5 MHz ELT provides only regional coverage.

- The position accuracy of the 406 MHz model provides precision within five kilometres (and up to 100 metres) where 121.5 MHz precision is within twenty kilometres.

- Search and rescue alert time is within five minutes for 406 MHz and an average of forty-five minutes with 121.5 MHz (COSPAS SARSAT 2008).

- The number of false alarms with 406 MHz ELTs is much lower than with 121.5 MHz models.

It is expected that the improvements offered in 406 MHz ELTs will result in dramatically improved search and rescue (SAR) response times. This is critically important because survival rates are highest in accidents with the quickest SAR response times. Ultimately, survival rates are probably less than 10 percent seventy-two hours after a crash (Transport Canada 2008).

Transport Canada agreed to ICAO's Standards and Recommended Practices (SARPs) to retrofit all aircraft with 406 MHz ELT's. However, the Federal Aviation Administration (FAA) in the United States has refused. This refusal was primarily based on strong opposition from the American Owners and Pilot's Association (AOPA) who argued that the 406 MHz ELT represents a significant cost and is beyond the needs of the typical GA owner-pilot. In the United States, GA represents 70 percent of the country's registered aircraft.

However, complications arise when countries choose not to comply with SARPS. As of 1 February 2009, American aircraft operating without a 406 MHz ELT are not permitted in Canadian airspace. This represents a particular challenge for aircraft attempting to fly through Canada to Alaska. This is an example of how complications can arise when countries choose to opt out of ICAO recommendations.

TC reports to the Canadian Parliament and Canadian citizens through the Minister of Transportation, Infrastructure and Communities. TC is a large organization with over 4,700 employees distributed across its headquarters and five regional offices in Vancouver, Winnipeg, Toronto, Montreal, and Moncton. These offices represent TC's five regions, which are Pacific, Prairie and Northern, Ontario, Quebec, and Atlantic, respectively.

The civil aviation sector of TC designs and operates a comprehensive national program to serve Canadians. The program is supported by 1500 employees and consists of five basic activities:

1. Qualification of aeronautical products, individuals, and organizations

2. Oversight of the aviation system

3. Education, promotion, and evaluation

4. Leadership and management

5. Safety policy, rulemaking, and agreements

The civil aviation sector of TC develops regulations for various aspects of the aviation industry including:

Airports

Aviation cannot exist without the network of airports across Canada. Transport Canada is responsible for establishing and enforcing airport safety and security standards. In addition, TC certifies and regulates all Canadian airports.

Aircraft

A large fleet of civil aircraft is active in Canada. TC's responsibility is to ensure that all civil aircraft conform to national and international standards related to maintenance, certification, and aircraft airworthiness. In addition, all new aircraft must be certified by TC before being sold to the public. New aircraft must meet very rigorous structural and performance standards before TC will grant certification.

Airlines and Aviation Operators

Canada's major scheduled and charter airlines must be certified and overseen by TC. This oversight is accomplished through TC audits, inspections and surveillance, validation, and assessment of safety management systems (SMS).

Safety

The aviation safety record in Canada is among the best in the world. TC strives towards maximizing air safety in Canada. This is accomplished through a proactive and integrated approach to prevent accidents before they occur. TC safety initiatives include the regulatory framework that promotes safety, safety oversight, airport funding through the Airports Capital Assistance Program, and aircraft services.

Security

TC is continually working towards maximizing the security of Canadian aviation. TC's security programs include the aviation security regulatory framework program, aviation security oversight program, airport policing assistance program, and the air cargo security initiative.

Emergencies and Incident Reporting

TC gathers information on incidents that involve aircraft registered in Canada or occurring at Canadian airports or within Canadian airspace. TC's Civil Aviation Contingency Operations Division is responsible for the organization's emergency preparedness for civil aviation. However, the Transportation Safety Board (TSB) of Canada is the agency responsible for aviation accident and incident data and investigation.

Environment

TC works directly with partners and stakeholders in aviation (such as the public, aviation industry, other government departments, provinces, territories, and municipalities, and international groups) to identify and reduce or prevent negative environmental impacts from aviation transportation. This work is conducted in conjunction with the International Civil Aviation Organization's Committee on Aviation Environmental Protection (CAEP) which is concerned with aircraft noise, engine emissions, and land use planning.

Licensing and Training of Aviation Personnel

TC is responsible for every element of pilot and flight engineer licensing in Canada. TC regulates flight schools, dictates the training requirements for each license and rating, administers flight crew examinations, and distributes licenses.

TC has an international reputation as a forward-thinking regulator that is highly-respected. In addition, being a smaller country, TC is less encumbered and can sometimes accomplish objectives easier or faster than larger States (such as the United States or Europe). For example, TC was among the first regulatory groups in the world to begin implementation of Safety Management Systems (SMS).

PROFESSIONAL PROFILE

Job Title:

Civil Aviation Safety Inspector at Transport Canada

Salary Range:

$50,000–90,000/year

Experience and Skills:

Safety inspectors typically have substantial industry experience. Safety inspectors come from a variety of backgrounds, including engineer, aircraft maintenance, flight attendant, lawyer, pilot, and air traffic control. TC looks for people who are familiar with and can work in a "systems" environment. A "systems" environment means that errors and mistakes are viewed as an inherent part of the system that can be identified, tracked, evaluated, and eventually reduced. Additional information on the systems approach to an organization can be found in Chapter 10.

Safety inspectors have strong communication skills, oral and written, as well as the ability to actively listen. This is crucial because interviews are an important part of the job. A high level of organizational knowledge is also required (for example, a familiarity with ICAO, the Federal Aviation Administration, and how the aviation industry works globally).

Responsibilities and Lifestyle:

The primary responsibility of safety inspectors is to ensure that aviation organizations are using safe practices and effectively running a Safety Management System (SMS).

The lifestyle of a regulator is a major advantage compared with operational jobs at an airline or flight school. For example, regulators do not have to live with a continual fear of layoffs, impossible hours, or long-stretches away from home. In addition, many regard the work as a public servant as being honourable, and most TC employees are motivated by serving the public.

Transportation Appeal Tribunal of Canada

The Transportation Appeal Tribunal of Canada was created to hold review and appeal hearings, when requested by interested persons, to look into administrative actions taken by the Minister of Transport. For example, if a person has been given notice by the Minister of Transport of the refusal to amend, suspension, or cancellation of their Canadian aviation document or if they have been assessed a monetary fine, they have the right to appeal that decision with the Tribunal.

The Tribunal is an independent quasi-judicial body, led by a full-time chairperson and vice-chairperson. Additionally, the Tribunal consists of full- or part-time members who have various types of aviation expertise. The Tribunal is conducted in an open and impartial manner, based on procedural fairness and the rules of natural justice. The Tribunal was established on 1 June 1986, resulting from recommendations made by Justice Charles Dubin in his *Inquiry into Aviation Safety in Canada* (Minister of Civil Aviation Tribunal 1997). Through the Tribunal, any member of the aviation community has the opportunity to have decisions made by the Minister of Transport reviewed by an independent body.

Prior to the creation of the Tribunal, TC could be thought of as the judge, jury, and executioner regarding aviation violations. If TC determined that you had contravened an air regulation, they issued a violation notice, assessed the penalty, and there was no appeal because it wasn't a charge under the criminal code. This process would not withstand a challenge from the Charter of Rights. The Tribunal allows for the review of TC decisions by an independent group.

Tribunal members are selected from within the aviation community. For example, Tribunal members may be Air Canada pilots, operators, and/or people with aviation experience. The Tribunal typically sits two to three times a month and will travel to the local communities to hear cases in that particular area.

The Tribunal is a quasi-judicial organization. This means that the person hearing the case isn't a judge; however, he or she will do basically the same thing as a judge. The person hearing the case has the right to take whatever action he or she thinks is proper, including dismissing the charge, changing the penalty, or even increasing the penalty.

There are two levels of Tribunal hearings. The first level is conducted by a single member of the Tribunal, in which he or she will review the case and make a determination. The second level of hearing, conducted by a designated chairperson and two Tribunal members, is an appeal of the determination made by the Tribunal member at the first level. Appeals are based on evidence, arguments, and exhibits in the transcript of the review hearing. Only new evidence, which was not previously available, may be presented at an appeal.

At an appeal, the Tribunal will sit as an appellate court to review the decisions that were made and make a final determination. Beyond the appeal, if a person is still unhappy with the decision, it is difficult to get into the court system because there is a privative clause that says that aviation matters stay with the Tribunal. However, if there is a jurisdictional issue, one can seek judicial review in the federal court.

The one limitation of the Tribunal is that it cannot make a medical decision. So, if TC refuses an aviator's medical certificate, he or she can

---

TO APPLY FOR A REVIEW HEARING:

A person who has received a notice from the Minister of Transport and wishes to present arguments to the Tribunal must make a written request for a review on or before the date specified in the notice of violation. This request should be sent to the Transportation Appeal Tribunal Registry at the following address:

The Transportation Appeal Tribunal of Canada
333 Laurier Avenue West
Room 1201
Ottawa, Ontario
K1A 0N5
fax (613) 990-9153
e-mail: info@tatc.gc.ca

The application should contain the following information:

- Name, address and telephone numbers
- Copy of the notice of violation received from the Agency
- Name and telephone number of representative, if any
- Hearing to be conducted in English or French

appeal the decision, but all that the Tribunal can do is send it back to the Minister of Transport for reconsideration. The Tribunal cannot reinstate a medical certificate.

Aviation Law

A crucially important responsibility of Transport Canada is the creation or modification, and enforcement of air laws. Broadly, laws can be broken down into two categories: Acts and Regulations. Acts are laws that are introduced through the legislative assembly. Regulations are subordinate legislations that are authorized by an Act and made by the Minister of Transport or a council.

Legislative Process (*Aeronautics Act*)

Before an Act comes into force, a bill must pass through the House of Commons and the Senate. For aviation acts, this process is initiated by the Minister of Transport who presents a bill for consideration and debate. In total, a bill will be reviewed three times through the House of Commons and the Senate before receiving Royal Assent and coming into force.

Air law within Canada is based on the *Aeronautics Act* of 1985 which authorizes control over aeronautics within Canada. The *Aeronautics Act* says that the Governor General and Federal Cabinet, and effectively the Minister of Transport, are responsible for developing and regulating aeronautics including aerodromes, aeronautical research, aircraft and equipment, aerial routes, and aviation safety. The Minister of Transport, in addition to authority over aviation, also has responsibility for all transportation services, including marine, rail, and road. However, there are some exceptions to this rule:

CANADIAN AVIATION REGULATION 102.01 STATES THAT THE FOLLOWING CATEGORIES DO NOT FALL UNDER THE AUTHORITY OF THE MINISTER OF TRANSPORT:

(A) MILITARY AIRCRAFT OF HER MAJESTY IN RIGHT OF CANADA WHEN THEY ARE BEING MANOEUVRED UNDER THE AUTHORITY OF THE MINISTER OF NATIONAL DEFENCE;

(B) MILITARY AIRCRAFT OF A COUNTRY OTHER THAN CANADA, TO THE EXTENT THAT THE MINISTER OF NATIONAL DEFENCE HAS EXEMPTED THEM FROM THE APPLICATION OF THESE REGULATIONS PURSUANT TO SUBSECTION 5.9(2) OF THE ACT; OR

(C) MODEL AIRCRAFT, ROCKETS, HOVERCRAFT OR WING-IN-GROUND-EFFECT MACHINES, UNLESS OTHERWISE INDICATED IN THE REGULATIONS.

The Supreme Court has determined that jurisdiction over aviation in Canada is reserved exclusively to the federal Parliament. Therefore, the provinces and municipalities have no right to legislate anything related to aeronautics, as it is under federal jurisdiction. This has interesting effects, because it results in municipal bylaws not applying to airports or aerodromes. For example, a person would not have to apply for a building permit to build a hangar on an aerodrome because that's a requirement of a municipal bylaw. In addition, the property upon which an airport is built does not have to be zoned for an airport because zoning bylaws are another municipal statute. Only the Federal Minister of Transportation can regulate the location and operation of airports and aerodromes.

Regulatory Process (*Aviation Regulations*)

The goal of the regulatory process is to produce regulations that result in the "greatest overall benefit to current and future generations of Canadians" (Government of Canada 2007 1). Regulations are a form of law, having binding legal effect. Air regulations are made by the Minister of Transport, to which Parliament has delegated authority over aviation.

Canadian Aviation Regulations (CARs)

Before 1996, regulations were scattered in a confusing and complicated system. Although most of the same regulations existed that are in effect today, they were difficult to find and were scattered within standards, schedules, and regulations. After 1996, regulations were ratified into the Canadian Aviation Regulations (CARs) which are much more accessible and understandable. This revision was based on recommendations from the Moshansky Commission, Transportation Safety Board, 1992–3 Federal Regulatory Review, and the TCA Rules Harmonization Project.

Canadian Aviation Regulations (CARs) are a collection of regulatory requirements that have been crafted to enhance the safety and competitiveness of aviation in Canada. CARs are commonly referred to as "air law" and describe the rules that those within civil aviation in Canada must abide by. CARs are accessible online through the TC website. The CARS are comprised of eight parts, as described by Transport Canada (2007):

PART 1—GENERAL PROVISIONS: DEFINITIONS, GENERAL ADMINISTRATIVE AND COMPLIANCE PROVISIONS, REGULATORY AUTHORITIES AND FEES FOR SERVICES PROVIDED BY THE DEPARTMENT.

PART 2—IDENTIFICATION, REGISTRATION AND LEASING OF AIRCRAFT: REGULATES REGISTRATION, MARKING AND LEASING OF AIRCRAFT AND IDENTIFICATION OF AERONAUTICAL PRODUCTS.

PART 3—AERODROME AND AIRPORTS: REGULATIONS RESPECTING AERODROMES AND AIRPORTS, AND REQUIREMENTS FOR CERTIFICATION OF AIRPORTS.

PART 4—PERSONNEL LICENSING AND TRAINING: REGULATIONS GOVERNING THE TRAINING AND LICENSING OF FLIGHT CREW, AIRCRAFT MAINTENANCE ENGINEERS AND AIR TRAFFIC CONTROLLERS.

PART 5—AIRWORTHINESS: REGULATES AIRWORTHINESS OF AIRCRAFT FROM THE DESIGN AND TYPE CERTIFICATION STAGE TO THE MAINTENANCE OF AIRCRAFT IN USE. INCLUDES REQUIREMENTS RESPECTING EXPORT, MANUFACTURE, AND DISTRIBUTION OF AIRCRAFT AND AERONAUTICAL PRODUCTS, AND REQUIREMENTS RESPECTING CONTINUING AIRWORTHINESS.

PART 6—GENERAL OPERATING AND FLIGHT RULES: GENERAL RULES APPLICABLE TO ALL AIRCRAFT OPERATIONS, INCLUDING REGULATIONS RESPECTING SPECIAL TYPES OF OPERATIONS SUCH AS AIR SHOWS, PARACHUTING AND BALLOON OPERATIONS.

PART 7—COMMERCIAL AIR SERVICES: RULES GOVERNING THE USE OF AIRPLANES AND HELICOPTERS IN COMMERCIAL AIR SERVICES, INCLUDING AIRWORTHINESS RULES RELATING SPECIFICALLY TO COMMERCIAL OPERATIONS. REFLECTS THE EVOLUTION OF THE AVIATION INDUSTRY IN CANADA WITH RESPECT TO OPERATIONS SUCH AS AERIAL WORK, AIR TAXI AND COMMUTER OPERATIONS. ALSO TAKES INTO ACCOUNT THE WAY COMMERCIAL AIR SERVICE REGULATIONS ARE STRUCTURED INTERNATIONALLY.

PART 8—AIR NAVIGATION SERVICES: REGULATIONS RESPECTING THE PROVISION OF AIR NAVIGATION SERVICES.

Canadian Aviation Regulation Advisory Council (CARAC)

In addition to the CARs, another aspect of the regulatory system which was introduced in 1996 is the Canadian Aviation Regulation Advisory Council (CARAC). CARAC is a committee made up of Transport Canada officials and members of a number of external groups who represent the interests of the aviation community. The CARAC committee includes representatives from management and labour organizations, operators, manufacturers, and professional organizations (Transport Canada 2007). Through cooperation, CARAC is able to broadly represent the interests of the aviation community.

IS A CHANGE NEEDED TO THE CARs?

As a member of the aviation community you have a right to request a regulatory change. The following information is required when requesting a regulatory change proposal:

- **Requests for Regulatory Action**
- **Technical Committee Agenda Item**

File Number	
Regulatory Reference	
Subject Title	
Sponsor or Petitioner (indicate OPI Branch if sponsored by TC)	

- **Description:**
- **Justification for Change:**
- **Current Regulatory text:**
- **Text of Proposed Regulatory Change:**

The purpose of CARAC is to "assess and recommend potential regulatory changes through cooperative rule-making" (Transport Canada 2007 ¶2). If any member of the aviation community wants a regulation changed or modified they are able to present their arguments to CARAC. Transport Canada is very receptive to worthwhile ideas from the industry or public at large. Therefore, members of the aviation community should never feel powerless when dealing with regulations. If there is something that needs to be changed and there's a reason for it, one should become involved in the CARAC process and help make the changes.

CARAC's activities are accomplished through various technical committees which provide advice and recommendations on proposals for regulatory amendment to the Civil Aviation Regulatory Committee (CARC) regarding the range of TC's regulatory mandate. There are ten technical committees with different focuses including:

- General
- Fees
- Identification, registration & leasing of aircraft
- Aerodromes and airports

- Personnel licensing and training
- Aircraft certification
- Maintenance and manufacturing
- General operating and flight rules
- Commercial air service operations
- Air navigation services and airspace

CARC is composed of senior TC Civil Aviation Safety executives who provide recommendations to the TC Assistant Deputy Minister, Safety and Security. Once CARC approves the recommendations of the committees, instructions are sent to the Department of Justice where lawyers prepare draft rules for transmission to the Minister of Transport for approval (called "discussion drafts"). Once the Minister of Transport approves, the file is sent to the Treasury Board for approval, and, once approved at this stage, it goes through the *Canada Gazette* process.

CARAC allows TC the opportunity to exchange ideas, information, and insight with industry representatives regarding new or modified regulations. This results in the creation of better regulations in less time.

Creating or Modifying Regulations

The stimulus that initiates the CARAC process comes from external sources, such as the aviation industry or ICAO, or from issues that impact aviation, including safety or environmental factors (refer to Figure 5.1). Upon evaluation of the stimulus, a decision will be made regarding whether a regulatory or non-regulatory alternative is required. An example of a Regulatory alternative would be an amendment to the CARs or associated standards that are enabled by the CARs. An example of a non-regulatory alternative would be guidance material or information booklets.

If a regulatory response is required, the CARAC process will begin, and tasks will be assigned to the appropriate technical committee.

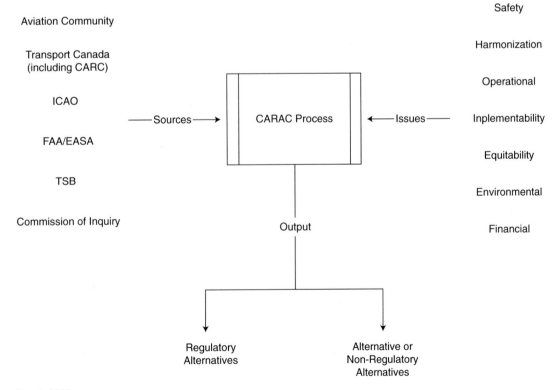

Transport Canada 2008.

Figure 5.4 *CARAC Process Stimulus*

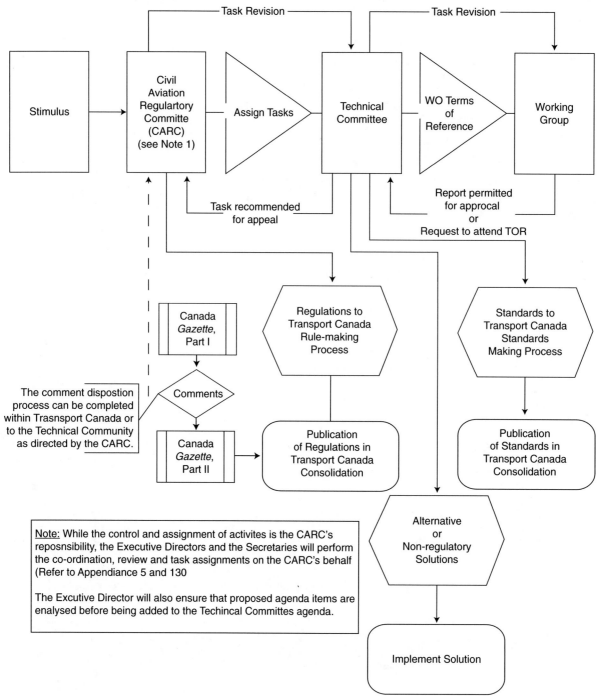

Transport Canada 2008.

Figure 5.5 *CARAC Process Workflow*

The CARAC workflow process is illustrated in Figure 5.5. From the initial stimulus, the CARAC process will work through various channels until an appropriate response is decided on. There are three categories of outcomes, resulting from the CARAC process: non-regulatory solutions (such as an information booklet), a standard, or a regulation.

The difference between a standard and a regulation is that a regulation is a rule of conduct. An associated standard is more technical in nature and should generally contain the application criteria in order to enable compliance with the rule (the "how to"). Standards are enforceable when they are enabled by the regulation (incorporated by reference within the regulation). CARAC is the consultation forum for standards although regulations are also presented for consultation at CARAC.

The *Canada Gazette* process is the formal consultation forum for regulations. Standards are not presented for consultation through the *Canada Gazette* process. When approved by a CARC executive, standards go through a legal review exercise and are published directly into the CARs.

The Regulatory Process

When the CARAC process determines that a regulation is required, the regulatory process begins. The regulatory process in Canada requires Transport Canada, in consultation with Regulatory Affairs, to assess proposals for new regulations early in the regulatory process. There are several factors considered during the assessment including:

- **POTENTIAL IMPACT OF THE REGULATION ON HEALTH AND SAFETY, SECURITY, THE ENVIRONMENT, AND THE SOCIAL AND ECONOMIC WELL-BEING OF CANADIANS;**

- **COST OR SAVINGS TO GOVERNMENT, BUSINESS, OR CANADIANS AND THE POTENTIAL IMPACT ON THE CANADIAN ECONOMY AND ITS INTERNATIONAL COMPETITIVENESS;**

- **POTENTIAL IMPACT ON OTHER FEDERAL DEPARTMENTS OR AGENCIES, OTHER GOVERNMENTS IN CANADA, OR ON CANADA'S FOREIGN AFFAIRS; AND**

- **DEGREE OF INTEREST, CONTENTION, AND SUPPORT AMONG AFFECTED PARTIES AND CANADIANS (GOVERNMENT OF CANADA 2007).**

However, if there is an immediate risk to security, the economy, or the environment (such as to aviation security directly after the 9/11 incidents) the regulatory process may be expedited.

Once the assessment is completed, the regulations are not developed in isolation. TC will identify and work with members of the aviation community, inviting participation at all stages of the process. The official method of soliciting consultation for regulatory proposals is through publication in the *Canada Gazette*. After publication in the *Gazette*, a specified number of days (which is usually thirty for domestic regulation and seventy-five for regulations with an international impact) are permitted for the collection of comments.

THE *CANADA GAZETTE*

The *Canada Gazette* is the Government of Canada's official newspaper that has been published by the Queen's Printer since 1841. The *Gazette* is utilized to distribute new statutes and regulations, proposed regulations, decisions of administrative boards and an assortment of government notices to the general public.

The *Gazette*, Parts 1, 2, and 3, are available in most public libraries.

Following, TC will identify and assess public policy issues, including the potential risks of the regulation, based on knowledge and evidence available. Once this assessment is complete, and it has been determined that a new or modified regulation is required, TC will set objectives that address the public policy issue and its causes, establish linkages to enabling legislation and government priorities to ensure relevance and consistency, and develop and use performance indicators on an ongoing basis to monitor and report on progress against performance expectations.

From this point, TC will assess how effective and appropriate the regulations are at achieving the objectives, assess any legal implications, comply with international obligations regarding human rights, health, safety, security, trade, and the environment, coordinate and cooperate with international, federal and provincial governments, analyze the costs and benefits of the regulation, and recommend the option that maximizes net benefits.

Next, TC completes an implementation plan and a plan as to how to enforce and measure compliance with the regulation. Once a regulation is in practice, TC will make continual assessments to ensure that it meets policy objectives. This occurs through a series of measurements, evaluations, and reviews (Government of Canada 2007).

Conclusion

The nature of the aviation industry is that it is international. For any international industry to succeed, regulations must be in place and agreed on by all of the countries involved. ICAO is the regulatory body which has been working since the 1940s to ensure safe, secure, and sustainable aviation on an international level. ICAO is an active group that is continually exploring future trends and distributing recommended practices to educate the international aviation community of best practices. However, ICAO does not directly control aviation regulations.

Within Canada, Transport Canada, and effectively the Minister of Transport, has authority over air law. Transport Canada has an international reputation as a forward-thinking and safety-conscious regulator. CARAC is a section of Transport Canada which uses an open and cooperative process to create or modify Canadian Aviation Regulations.

It is important that members of the aviation community do not allow themselves to be overwhelmed by the regulatory system. When effort is invested into understanding the regulatory process, it is evident that it is an open and straight-forward process that invites feedback and interaction with the public on several levels. It is crucial that members of the aviation community become involved in the regulatory process to develop regulations and standards that help Transport Canada maintain Canada's position as a world leader in the aviation industry.

PRACTICE ACTIVITY

On 15 January 2009, just minutes after takeoff, US Airways flight 1549 ditched (emergency landed on water) in the Hudson River after both engines were hit by geese. This incident raised questions about how to reduce the likelihood of similar accidents in the future. One option is to reduce the speed of aircraft on departure, while below 10,000 feet above sea level, to less than 250 knots. The reasoning for this is that a 20 percent increase in airspeed, from 250 knots to 300 knots, results in a 44 percent increase in the force of the impact on the airframe of the aircraft. The consequences of bird strikes are most severe during the departure phase of flight (*Canada Gazette* 2009). A draft of a modified regulation is presented below:

CAR 602.32

(1) SUBJECT TO SUBSECTION (2), NO PERSON SHALL

    (A) OPERATE AN AIRCRAFT AT AN INDICATED AIRSPEED OF MORE THAN 250 KNOTS IF THE AIRCRAFT IS BELOW 10,000 FEET ASL; OR

    (B) OPERATE AN AIRCRAFT AT AN INDICATED AIRSPEED OF MORE THAN 200 KNOTS IF THE AIRCRAFT IS BELOW 3,000 FEET AGL WITHIN 10 NAUTICAL MILES OF A CONTROLLED AERODROME UNLESS AUTHORIZED TO DO SO IN AN AIR TRAFFIC CONTROL CLEARANCE.

(2) A PERSON MAY OPERATE AN AIRCRAFT AT AN INDICATED AIRSPEED GREATER THAN THE AIRSPEEDS REFERRED TO IN SUBSECTION (1) IF THE AIRCRAFT IS BEING OPERATED IN ACCORDANCE WITH A SPECIAL FLIGHT OPERATIONS CERTIFICATE—SPECIAL AVIATION EVENT ISSUED PURSUANT TO SECTION 603.02.

(3) IF THE MINIMUM SAFE AIRSPEED FOR THE FLIGHT CONFIGURATION OF AN AIRCRAFT IS GREATER THAN THE AIRSPEED REFERRED TO IN SUBSECTION (1), THE AIRCRAFT SHALL BE OPERATED AT THE MINIMUM SAFE AIRSPEED.

Pretend that you are a member of CARAC and are assessing this proposed regulation. Answer the following questions:

1. What are the costs and benefits of this change to stakeholders?

_____

_____

_____

_____

_____

_____

2. What will be the impacts to businesses and the travelling public?

_____

_____

_____

_____

_____

3. Is there a non-regulatory option that would adequately address this issue? Why or why not?

_____

_____

_____

_____

_____

4. How would this regulation be enforced (for example, monetary penalties)? Who will enforce it?

_____

_____

_____

_____

_____

CHAPTER **6**

AIR NAVIGATION SYSTEM OF CANADA

OBJECTIVES

After you have read this chapter, you will:

- Understand key regulations governing air navigation

- Be familiar with the organization and structure of airspace in Canada

- Know the history, organization, and mandate of NAV CANADA

- Be able to describe the roles, responsibilities, and equipment of air traffic controllers

- Know whether the pilot or air traffic controller ultimately has the safety authority over each flight

Introduction

Air navigation allows for the piloting of aircraft from departure to destination airports in a safe and expeditious manner. Just as all types of vehicles share a highway, all aircraft utilize the navigation system and must abide by the rules of air navigation.

Canada's air navigation system (ANS) is comprised of three broad elements:

- Infrastructure—airport infrastructure, airspace, and regulations that govern air navigation

- People—everyone involved in operating and utilizing the ANS (pilots, air traffic controllers, etc.)

- Equipment—aircraft that operate within the system, navigation aids, lighting, and control facilities

Without all of these elements working together, safe and efficient air navigation would be impossible.

Infrastructure

The infrastructure of Canada's air navigation system is comprised of airport equipment and regulations (including the organization of airspace). In Canada, airports are operated by various entities. For additional information on Canadian airports, review Chapter 4. Canadian regulations are written, established, distributed, and enforced by Transport Canada. More information about Canadian Aviation Regulations and Transport Canada can be found in Chapter 5. This section will describe components of the ANS infrastructure, specifically airport equipment and key regulations that govern navigation.

Airport Infrastructure

Each airport, regardless of size, is associated with important infrastructure to support air navigation. Examples of this infrastructure, at a minimum, include:

- Fuel services

- Maintenance of runway surfaces

- Runway and approach lighting

- Operational runways and taxiways

- Designated radio frequencies

- Approach and departure procedures (which range in complexity from a standard circuit pattern in use at small uncontrolled airports to published standard terminal arrival routes and standard instrument departures at larger airports)

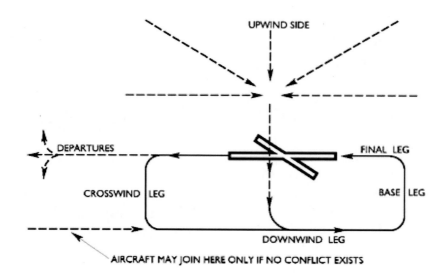

Figure 6.1 *Standard Circuit Pattern at an Uncontrolled Airport*

Regulations

Regulations represent another component of Canada's air navigation system infrastructure. Regulations that govern air travel in Canada are produced by Transport Canada (TC). To provide a foundational understanding of how regulations impact air navigation in Canada, visual- and instrument-flight-rules, airspace, and airways will be reviewed.

Visual Flight Rules (VFR) and Instrument Flight Rules (IFR)

Aircraft navigate under either visual flight rules (VFR) or instrument flight rules (IFR). Visual flight rules (VFR) flight is conducted with visual reference to the ground. This means that, in addition to referencing onboard instruments, the pilot relies heavily on watching the geographical characteristics of the land (lakes, hills, towns, etc.) and comparing them to his or her map to determine aircraft position. In order to conduct a VFR flight, Canadian Aviation Regulations (Division VI—602.114) specify minimum visibility and distance-from-clouds requirements. These requirements ensure that pilots can maintain visual contact with the ground and are able to watch for other aircraft that may pose a collision threat. Flight training, sightseeing, parachuting, crop-dusting, and traffic watch flights are typically conducted under VFR.

The only way a pilot is permitted to fly without visual reference to the ground, such as in fog or through clouds, is by following instrument flight rules (IFR). Although minimum visibility requirements are established for the takeoff and landing phases of IFR flights, IFR is conducted by relying almost exclusively on cockpit instruments. During IFR flights, air traffic controllers (ATC) are usually in contact with pilots. ATC will track the aircraft's position via a surveillance system or pilot position reports. Since IFR flight is frequently conducted in weather conditions with little visibility, pilots cannot be responsible for maintaining safe separation from other aircraft. Therefore, ATC is responsible for the safe separation of IFR aircraft from other traffic when the IFR traffic is in controlled airspace. Controllers will be in continual communication with pilots, providing clearances and instructions, to accomplish this task. Almost all commercial traffic, including all airline flights, is conducted under IFR.

Canadian Domestic Airspace—Classification and Structure

Classification

All of the airspace in Canada is classified into one of seven categories. Airspace is organized alphabetically from A to G. Specific operating rules, ATC service levels, communication requirements, and equipment requirements exist depending on what airspace an aircraft is flying in. Refer to Figure 6.2 for an overview of Canadian airspace. Before reviewing the alphabetic classes, it is important to understand the difference between controlled and uncontrolled airspace. In controlled airspace, ATC provides control services to pilots. In uncontrolled airspace ATC provides other services, but not control services. For example, in uncontrolled airspace, ATC will not separate one aircraft from another aircraft. However, workload permitting, ATC might provide traffic advisory information.

Class A

Class A airspace includes high-level controlled airspace from 18,000 feet above sea level (ASL) to flight level (FL) 600. FL 600 is roughly equivalent to 60,000 feet above sea level. Class A airspace is limited to IFR traffic.

Class B

Airspace is categorized as Class B when IFR service is required, in addition to the need to control VFR traffic. Low-level controlled airspace is above 12,500 feet ASL or from the minimum en route altitude (MEA), whichever is higher, to below 18,000 feet ASL. Terminal control areas (TCAs) and associated primary control zones (CZs) may also be classified Class B airspace. VFR aircraft must get a clearance from ATC before entering Class B airspace and maintain visual contact with the ground.

Class C

This classification of airspace is typically associated with the terminal control area (TCA) and control zone (CZ) of large airports. Both IFR and VFR flights are permitted in Class C airspace. VFR flights must obtain a clearance from ATC to enter Class C airspace. In Class C airspace, aircraft are provided with positive conflict resolution. If the ATC unit within the Class C airspace shuts down for any reason, the airspace converts to Class E.

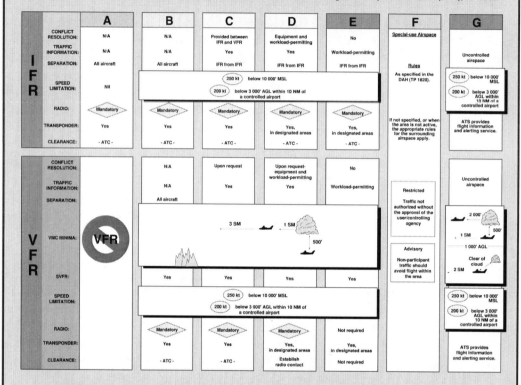

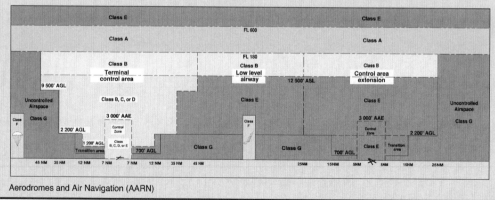

Transport Canada Aeronautical Information Manual TP 14371.

Figure 6.2 *Overview of Canadian Airspace*

Class D

Class D airspace is associated with terminal control areas (TCAs) and control zones (CZs) of smaller airports that have a control tower. Both IFR and VFR flights are permitted in Class D airspace. VFR flights do not require a clearance to enter the airspace, but they must establish radio communication with ATC. If the ATC unit within the Class D airspace shuts down for any reason, the airspace converts to Class E.

Class E

This class of airspace refers to other controlled airspace and is present in several forms. These include all high-level controlled airspace above FL 600, low-level airways, low-level fixed area navigation (RNAV) routes, control area extensions (CAEs), transition areas (TAs) or control zones (CZs) established without an operating control tower. IFR flights still require a clearance to operate in Class E and are provided with separation from other IFR flights. VFR aircraft require only the appropriate weather conditions to fly with visual reference to the ground. VFR flights are not required to communicate with ATC in Class E airspace.

Class F

The purpose of Class F airspace is to isolate activities that may threaten the safety of aircraft in the vicinity. This airspace is considered special-use and is classified as restricted, advisory, military operations, or danger areas, depending on the activity. This airspace may be controlled, uncontrolled, or a combination of both. Class F airspace is described in terms of horizontal and vertical dimensions, effective for a specified period of time.

Class G

This airspace is uncontrolled. It basically refers to all airspace in Canada that doesn't fall into one of the above categories. Clearances are not required for pilots to operate in this airspace. Class G airspace is found in remote areas with minimal traffic, as well as along a section of airspace close to the ground and in some areas between 12,500 feet and 18,000 feet. Airspace shall be classified G if it has not been designated A, B, C, D, E or F.

Structure

Control Area Extensions

Control area extensions (CAEs) are sections of airspace around airports where the controlled airspace alone would not contain IFR traffic and allow for adequate separation between IFR departures and arrivals. CAEs provide additional space around busy airports where controllers can offer IFR control and IFR aircraft have additional maneuvering space. In addition, CAEs often provide connective airspace between an airport's control zone and other airspace to allow aircraft to continue en route. CAEs are based at 2,200 feet above the ground and go up to, but not including, 18,000 feet above sea level.

Control Zones

In Figure 6.3, there is a picture of a cylindrical shaped control zone (CZ) overlying the airports. Most major Canadian airports are surrounded by a CZ. The purpose of CZs is to keep IFR aircraft in controlled airspace during approaches and to control VFR traffic. CZs start at surface level and surround the airport by a five or seven nautical mile radius. CZs typically ascend to 3000 feet above the elevation of the airport.

Transition Areas

Transition areas (TAs) are created to provide an additional section of controlled airspace to facilitate IFR operations. TAs are typically Class E airspace based at 700 feet above the ground and go up to the base of overlying controlled airspace. TAs are meant to reduce the number of close encounters between IFR aircraft on approach and VFR aircraft surrounding the CZ.

Terminal Control Areas

Terminal control areas (TCAs) are sections of airspace overlying busy airports. The purpose of a TCA is to offer additional IFR control service to en route, arriving, and departing aircraft. There are strict rules associated with TCAs that govern the communication, equipment, and procedure of each flight. These rules are based on what ATC determines to be appropriate for the traffic volume.

TCAs differ from control area extensions in that they may extend to high-level airspace, IFR traffic is usually controlled by a terminal control unit (TCU), and the TCA has a circular configuration.

It is important to note that as you get further away from the airport, the TCA steps up in a tier shape that is commonly described as an upside-down wedding cake. This allows for uncontrolled Class G airspace around an airport, where training flights or recreational aviation activities can occur, without interfering with the activities of IFR aircraft. The purpose of a TCA is to ensure that IFR traffic on approach to or departure from the airport have detailed information about traffic in the area. Legally, there will not be any unknown VFR traffic in a TCA. TCAs are meant to reduce close encounters between IFR aircraft and VFR aircraft surrounding the CZ.

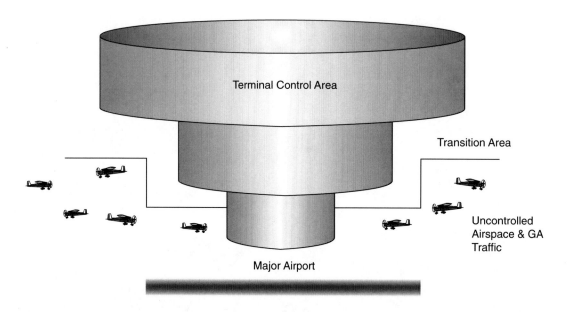

Figure 6.3 *TCA and surrounding GA traffic in uncontrolled airspace.*

People and Equipment

The people and equipment in the civilian air navigation system (ANS) are operated by a single entity—NAV CANADA. This section will explore the history, facilities, operation, and people that supply air traffic control services in Canada.

NAV CANADA

On 1 November 1996, Transport Canada transferred the responsibility for Canada's air navigation system (along with all Canadian air traffic control [ATC] facilities) to a private sector non-share capital company named NAV CANADA. This transaction cost $1.5 billion. This was an historic arrangement, as NAV CANADA was the first private sector company to commercialize a government service. Although NAV CANADA is a private company, safety performance is regulated by Transport Canada.

THE IMPORTANCE
OF MONITORING AIRSPACE

On 31 August 1986, Aeronaves de Mexico Flight 498 departed Mexico City headed for Los Angeles airport in the United States. This flight was conducted in a DC-9 aircraft. Fifty-eight passengers and six crew members were aboard. As a regularly scheduled passenger flight, this aircraft was flying under Instrument Flight Rules (IFR) and was in contact with Air Traffic Control. At 11:46, Flight 498 began its descent to the Los Angeles airport.

On the same day at 11:41, a small General Aviation (GA) Piper PA-28 aircraft took off from Torrance, California, headed towards Big Bear, California, with the pilot and two passengers aboard. Like many GA aircraft, this flight was conducted under Visual Flight Rules (VFR) and was not in contact with air traffic control. The pilot was relatively new to the area and, during the climb, the pilot inadvertently entered the Los Angeles airport's Terminal Control Area (TCA) without contacting ATC or receiving the required clearance.

At 11:52, the engine of the GA Piper contacted the left horizontal stabilizer of Flight 498 over Cerritos, California, at approximately 6,560 feet. The horizontal stabilizer cut through the cabin of the Piper, severing a large portion of the top of the cabin and decapitating all three people on the Piper aircraft. At the time of the collision, the sky was clear with fourteen miles visibility. This is an important point, because investigators were shocked that the aircraft did not see each other until it was too late. Investigation later revealed that visual scan techniques, when coupled with managing onboard systems, rarely allow pilots to notice conflicting traffic until it is too late to avoid a collision.

Sixty-four passengers on Flight 498 were killed, as were the three people on the GA Piper. The aircraft wreckage and fires damaged several houses and killed fifteen people on the ground.

Class G uncontrolled airspace often surrounds TCAs (refer to Figure 6.3). It is common for GA traffic to be operating around the TCA. The controller's priority is to separate IFR aircraft from other IFR aircraft. Flight 498 had equipment called a Mode C transponder on board that provided ATC with position and altitude information. The Piper had a transponder (not Mode C) that provided only position information. Without altitude information, the controller handling Flight 498 ignored the transponder reading from the GA aircraft and assumed that it was below the TCA airspace and not a threat to IFR aircraft within the TCA on approach or departure to the airport. "The entry of the Piper pilot into the TCA stripped his airplane and Flight 498 of the precise protection the TCA was designed to provide. Its entry into this prohibited airspace created an exposure to risk that should never have existed" (NTSB 1986 43).

NAV CANADA's mandate is to promote the safety and efficiency of Canada's ANS. Airlines, and other aircraft owners and operators, are provided with services including air traffic control (ATC), flight information, weather briefings, airport advisory services, aeronautical information, and ground-based electronic navigation aids.

Safety and efficiency represent the two goals of NAV CANADA. Safety is paramount because, if the flight does not arrive at the destination airport safely, all other concerns are irrelevant. However, in order for aviation companies to survive economically, efficiency is also crucial. Flight delays cost airlines an enormous amount of money in fuel, flight crew salaries, rerouting of passengers who miss connections, maintenance (as the aircraft will reach mandatory inspection times more quickly), and aircraft time. If Canada's air navigation system is inefficient, it makes the financial survival of aviation companies very difficult. Therefore, the two major goals of air transportation are safety and efficiency.

NAV CANADA was created through the collaborative efforts of employees, unions, pilots, airlines, government officials, and other members of the aviation sector who share an interest in the ability of the Canadian ANS to meet the challenges of the future. The collaboration in the creation of NAV CANADA was important to ensure the company had a fair and appropriate operational strategy. NAV CANADA's current corporate governance structure reflects its origins. The governing Board of Directors is comprised of representatives from several stakeholder aviation groups. The fifteen-member board includes representatives from air carriers, general and business aviation, the federal government, bargaining agents, and independents who are appointed by the board and have no ties to stakeholders. The president of NAV CANADA is also a director.

In 2007, NAV CANADA handled 11.6 million aircraft movements.

NAV CANADA has approximately 5,200 employees. In addition to ATC and flight service specialists, the company employs electronic technologists, operational support specialists, engineers, managers, and administrative staff.

NAV CANADA generates revenue through customer service charges to airlines and aircraft operators for navigation services. For small general aviation aircraft, a fee of $68 per year is charged to the aircraft's owner. Rates for larger aircraft vary and are dependent on the weight of the aircraft and the distance flown, calculated on a per passenger basis. NAV CANADA charges in 2008 were approximately 30 percent less than the Air Transportation Tax, which existed prior to the creation of the company. Since 1999, when NAV CANADA service charges were introduced, charges have increased by around 5 percent (approximately seventeen points below inflation). These charges are made on a cost recovery basis, as the company is permitted to collect revenues only to cover its costs associated with the provision of air navigation services, plus reasonable and prudent financial reserves. NAV CANADA operates on a break-even basis.

NAV CANADA now operates completely independently of government funding. The company's acquisition of the ANS, as well as its ongoing capital requirements, is financed with debt. In 2008, the company had 1.925 billion dollars in outstanding bonds and medium-term notes.

Canada's air navigation system is the second largest in the world. NAV CANADA facilities include seven area control centres (Vancouver, Edmonton, Winnipeg, Toronto, Montreal, Moncton, and Gander), forty-one control towers, fifty-nine flight service stations, eight flight information centres, thirty-eight maintenance centres, and fifty community aerodrome radio stations (CARS) providing weather information to Canada's North. These facilities utilize over 1,000 ground-based navigation aids and forty-five radar sites across Canada. In 2007, NAV CANADA handled 11.6 million aircraft movements (including takeoffs, landings, and over-flights).

NAV CANADA's head office, technical systems centre, and simulation centre are located in Ottawa, Ontario. Their training and conference centre is located in Cornwall, Ontario.

NAV CANADA's safety record is one of the best in the world and is continually improving. For example, in 2001 there was an average number of loss-of-separation of IFR-to-IFR aircraft of 1 per 100,000 movements. In 2007, this rate had been reduced to 0.75 per 100,000 movements.

ANS Modernization

NAV CANADA is continually taking steps to modernize and increase the efficiency of the ANS in Canada. Since the company took over the ANS in 1996, they have invested over $1.3 billion in new systems and technology. It has been estimated that customers save $100 million annually because of the more efficient operations provided by NAV CANADA technology and services. This has resulted in NAV CANADA's technology being among the best in the world. An example of this is in the application of text-based messaging over the North Atlantic. The North Atlantic is the busiest oceanic airspace in the world with approximately 1,000 flights per day. This text-based messaging allows ATC to communicate with pilots electronically (similar to sending a text message on a cell phone) thereby reducing voice communication and congestion on the radio frequency. As only one person is able to speak at a time on the radio, this technology is capable of greatly increasing the speed and efficiency of communication. This system is called the Controller–lot Datalink (CPDLC) and is provided by the world-leading Gander Automated Air Traffic System (GAATS) for oceanic ATC.

Air Traffic Control Services

To gain an understanding of how all of the NAV CANADA services interact, it is helpful to put yourself in the position of a pilot about to commence an IFR flight at a large international airport. During the flight preparation, taxi, takeoff, departure, cruise, approach, landing, and final taxi to the gate the pilot will communicate with different NAV CANADA employees for different purposes. The following section will describe step-by-step the interaction a pilot would have with each group.

Flight Preparation

Before general aviation pilots walk to their aircraft, they must first contact a Flight Service Specialist via telephone at 1-866-WX-BRIEF (for English services) or 1-866-GO-METEO (for French services) to get information regarding the current and expected weather conditions, status of navigational aids, and whether there are any current airport advisories or emergency alerts. Another important role of FSS is in the filing of flight plans. Before conducting a non-local flight, pilots must file their intentions in a flight plan. Flight plans contain detailed information about the:

- Aircraft
- Flight rules chosen (IFR or VFR)
- Equipment onboard
- Departure airport
- Time of departure
- Speed
- Altitude
- Route
- Destination airport
- Estimated time en route
- Time after which authorities should contact search-and-rescue
- Number of people on board
- Survival equipment
- Name and address of pilot
- Name and contact information for a person or company to be contacted if search-and-rescue is initiated

Pilots can also contact a flight service specialist in-flight on the aircraft's radio to request updated weather information or to amend their flight plan. This service is provided to pilots through eight flight information centres (FIC). Traditionally, all flight service activities were provided by the fifty-nine flight service stations (FSSs), with each designed to provide a full range of aviation information services. However, as the industry evolved and improvements were required to the flight information services network, in 1998 NAV CANADA commissioned a study to investigate how to improve these services. Based on the results of this study, NAV CANADA developed FICs to consolidate flight information services.

Presently, FSSs are located at fifty-nine airports across Canada and provide specialized services, including local weather observations, airport advisory services, vehicle control, and assistance to aircraft in emergency situations. The eight FICs provide information that is not location dependent. These centralized flight information services include interpretive weather briefings, flight planning services, and en route advisories. Using the toll-free number, an aviator is automatically connected to the FIC in their area. FICs provide one-stop shopping for flight planning and weather briefings by specialists who use the latest technology.

It is important to note that airline pilots do not typically contact a FSS for a weather briefing or to file a flight plan. At the commercial level, this work is completed for the pilot by the airline dispatcher.

Flight Progress Strips

Before an aircraft begins taxiing, a flight progress strip (FPS) is prepared by ATC. An FPS is a small slip of paper that contains crucial information that controllers need. FPSs will be handed physically or, more commonly, electronically to every controller that is working with a flight. FPSs often serve as a physical reminder to controllers of all of the aircraft that they are working with at a given point in time. As a flight progresses,

NAV CANADA

CANADIAN FLIGHT PLAN AND FLIGHT ITINERARY
PLAN DE VOL ET ITINÉRAIRE DE VOL CANADIEN

ICAO FLIGHT PLAN
PLAN DE VOL OACI

PRIORITY / PRIORITÉ   ADDRESSEE(S) / DESTINATAIRE(S)

<< ≡ **FF** →

<< ≡

FILING TIME / HEURE DE DÉPÔT   →   ORIGINATOR / EXPÉDITEUR   << ≡

SPECIFIC IDENTIFICATION OF ADDRESSEE(S) AND/OR ORIGINATOR / IDENTIFICATION PRÉCISE DU(DES) DESTINATAIRE(S) ET/OU DE L'EXPÉDITEUR

3 MESSAGE TYPE
   TYPE DE MESSAGE

<< ≡ **(FPL**

7 AIRCRAFT IDENTIFICATION /
   IDENTIFICATION DE L'AÉRONEF

8 FLIGHT RULES /
   RÈGLES DE VOL

TYPE OF FLIGHT /
TYPE DE VOL

9 NUMBER / NOMBRE   TYPE OF AIRCRAFT / TYPE

WAKE TURBULENCE CAT. /
CAT. DE TURBULENCE DE   /

10 EQUIPMENT / ÉQUIPEMENT   <<

13 DEPARTURE AERODROME / AÉRODROME DE DÉPART   TIME / HEURE   << ≡

15 CRUISING SPEED /
    VITESSE DE CROISIÈRE   ALTITUDE / LEVEL / NIVEAU   ROUTE / ROUTE

0 0 0 0 0   0 0 0 0 0 →

<< ≡

16 DESTINATION AERODROME
    AÉRODROME DE DESTINATION   TOTAL EET / DURÉE TOTALE ESTIMÉE
    DAYS/JOURS   HRS   MINS   SAR
    HRS   MINS   ALTN AERODROME /
    AÉRODROME DE DÉGAGEMENT   2ND ALTN AERODROME /
    2e AÉRODROME DE DÉGAGEMENT   << ≡

18 **OTHER INFORMATION / RENSEIGNEMENTS DIVERS**

)<<

19   ENDURANCE / AUTONOMIE   EMERGENCY RADIO / RADIO DE SECOURS

    HRS   MINS   PERSONS ON BOARD / PERSONNES À BORD   UHF   VHF   ELT   ELT TYPE / TYPE D'ELT

E /   → **P /**   → **R/ U**   **V**   **E**

SURVIVAL EQUIPMENT / ÉQUIPEMENT DE SURVIE   **JACKETS / GILETS DE SAUVETAGE**

    POLAR
    POLAIRE   DESERT
    DÉSERT   MARITIME
    MARITIME   JUNGLE
    JUNGLE   LIGHT
    LAMPES   FLUORES
    FLUORES   UHF   VHF

→ **S** / **P**   **D**   **M**   **J**   → **J** / **L**   **F**   **U**   **V**

DINGHIES / CANOTS

    NUMBER
    NOMBRE   CAPACITY
    CAPACITÉ   COVER
    COUVERTURE   COLOUR
    COULEUR

→ **D** /   →   → **C** →   << ≡

AIRCRAFT COLOUR AND MARKINGS / COULEUR ET MARQUES DE L'AÉRONEF   WHEELS
ROUES   SEAPLANE
HYDRAVION   SKIS   AMPHIBIAN
AMPHIBIE

A /

REMARKS / REMARQUES

→ **N /**   << ≡

AN ARRIVAL REPORT WILL BE FILED WITH / UN COMPTE RENDU D'ARRIVÉE SERA NOTIFIÉ À :

NAME AND PHONE NUMBER OR ADDRESS OF PERSONS(S) OR COMPANY TO BE NOTIFIED IF SEARCH AND RESCUE ACTION INITIATED /
NOM ET NUMÉRO DE TÉLÉPHONE OU ADRESSE DE LA (DES) PERSONNE(S) OU COMPAGNIE À AVISER SI DES RECHERCHES SONT ENTREPRISES

PILOT-IN-COMMAND / PILOTE COMMANDANT DE BORD   PILOT'S LICENCE NO. / N° DE LICENCE DU PILOTE

C /   )<< ≡

FILED BY / DÉPOSÉ PAR   SPACE RESERVED FOR ADDITIONAL REQUIREMENTS / ESPACE RÉSERVÉ À DES FINS SUPPLÉMENTAIRES

NAVCAN26-0516 (2004-01)

Many airports across Canada are uncontrolled, meaning that there is no air traffic control facility on the field. In this situation, standardized infrastructure is required to allow pilots to approach, sequence for landing, and land in a safe and efficient manner. This is most often accomplished through an established circuit pattern.

continues

THE CIRCUIT PATTERN (*CONTINUED*)

Circuit patterns are based off of which runway is active. The active runway is that which is most closely aligned with the wind—as aircraft always try to take off and land facing into the wind. If a pilot is unable to determine which runway is active based on radio communications with a ground facility or other aircraft, he or she may overfly the airport 500 feet above circuit altitude to visually check the airport windsock.

The circuit pattern is an oval-shaped route which is flown at 1000 feet above the ground. Pilots will join the circuit straight-in on the downwind leg if no traffic exists. Alternatively, pilots may join the circuit from the upwind wide of the airport. In this case, the pilot will be established at circuit altitude (typically 1000 feet above the ground) when over the runway and join a mid-downwind leg. The pilot will descend to 500 feet above the ground while on the base leg of the circuit and descend to landing during the final leg. This pattern adds a measure of standardization and safety because a pilot, even if unfamiliar with an airport, has an expectation as to where the traffic will be and how to land safely.

controllers will make notes on the strip to remind them of status changes. The benefit of flight progress strips is that controllers have advanced notice of what an aircraft's intentions are, thus eliminating the need for pilots to repeat information to every new controller they are in touch with.

A sample FPS is pictured below:

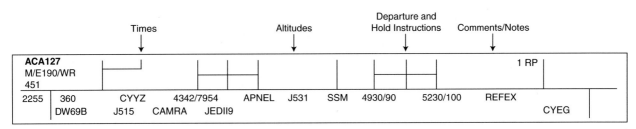

NAV Canada, 2009.

Figure 6.5 *Example of Flight Progress Strip*

From the example FPS above, the information is interpreted as follows:

ACA 127 = Air Canada flight 127

M = Medium weight class

E190 = Aircraft type is an Embraer 190.

WR = Transponder and navigational type

451 = True airspeed in nautical miles per hour

2255 = Transponder code

360 = Requested altitude of 36,000 feet above sea level

CYYZ = Departing Toronto International Airport

4342/7954 = Latitude and longitude

4342/7954–JEDII9 = Filed and cleared routing

CYEG = Destination airport is Edmonton International Airport.

Controllers often write notes on the strips to keep track of updated information. These notes are written on specific areas of the strip. The second box in the top row is where controllers will write information about times. The next large box to the right is where controllers will write information about altitudes. The next section of four small boxes is where controllers will indicate departure and hold instructions. The next large box is where controllers will write comments or notes.

As a component of its modernization initiatives, NAV CANADA has developed a new technology known as EXCDS (extended computer display system). This technology eliminates the need for paper FPS and allows controllers to manage flight data through a computer. This system has been operational in Canada since 1998 and is used at thirty-one air navigation facilities across Canada, including all seven ACCs and twenty-one control towers. NAV CANADA has sold the EXCDS system to many ATC agencies around the world, including to London's Heathrow airport, which is one of the busiest airport in the world.

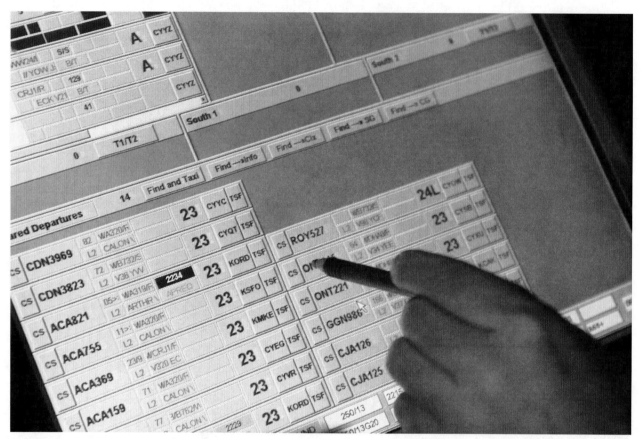

Copyright NAV CANADA 2009.

Figure 6.4 *Touch-screen Flight Strips (EXCDS)*

ATIS

Once pilots have completed pre-flight briefings and have filed their flight plans, they will walk to their aircraft and begin the process to initiate takeoff. Typically, the first radio frequency that a pilot will listen to is the airport's ATIS frequency. ATIS stands for Automatic Terminal Information Service. The ATIS is a pre-recorded message broadcast on a continual loop that provides information about the airport and current weather conditions.

Controllers expect that pilots have listened to ATIS information, and it is courteous and professional to do so before contacting ATC. The ATIS message is updated hourly or whenever conditions change. To identify that a message is new, each ATIS is coded with a letter from the phonetic alphabet. The phonetic alphabet replaces

PHONETIC ALPHABET	
A—Alfa	N—November
B—Bravo	O—Oscar
C—Charlie	P—Papa
D—Delta	Q—Quebec
E—Echo	R—Romeo
F—Foxtrot	S—Sierra
G—Golf	T—Tango
H—Hotel	U—Uniform
I—India	V—Victor
J—Juliett	W—Whiskey
K—Kilo	X—X-ray
L—Lima	Y—Yankee
M—Mike	Z—Zulu
	Source: ICAO Annex 10 Volume II

letters with standardized code words. This system is used throughout aviation to standardize communication and avoid confusion over letters that sound the same and might be confused, such as B, D, P, T.

The following is a sample ATIS broadcast from Toronto's L.B. Pearson International Airport:

"Toronto International information CHARLIE, weather at 1300 Zulu, wind 260 at 8, visibility 25, ceiling 8,000 scattered, temperature 24, dewpoint 12, altimeter 30 10. The IFR approach is ILS runway 24 right tower, frequency 118.35, and ILS runway 23, tower frequency 118.7. Departure runways are 24 right and 23. Inform Toronto ATC on initial contact that you have received information CHARLIE."

ATIS Text	Description
Toronto International information CHARLIE	This is ATIS information for the Toronto International Airport. "Charlie" is the identification letter of the broadcast. If ATIS is updated between the times a pilot listens to it and contacts ATC, he or she will be advised to listen again for the new current information "Delta."
Weather at 1300 Zulu	In aviation, as aircraft may cross several time zones during a flight, standard time must be used so that all airplanes are using the same clock. This is called several names, including Universal Coordinated Time (UTC), Greenwich Mean Time (GMT), or "Zulu." In Toronto, local time is −5 hours Zulu. Therefore, local time when the ATIS was recorded was 08:00.

Wind 260 at 8	The wind at the airport is blowing out of 260 degrees magnetic, roughly from a western orientation, at a speed of 8 nautical miles per hour. One nautical mile equals 1.852 kilometres.
Visibility 25	The visibility at the airport is 25 statute miles.
Ceiling 8,000 scattered	A scattered layer of clouds is present at 8,000 feet above airport elevation.
Temperature 24, dewpoint 12	The temperature at the airport is 24 degrees Celsius, and the dewpoint is 12 degrees Celsius.
Altimeter 30 10	The altimeter setting at the airport is 30.10. Most altimeters in aircraft measure altitude based on the decrease in atmospheric pressure as the plane ascends. However, the atmospheric pressure at ground level varies, based on weather conditions. Therefore, in order to ensure an altimeter is accurate, the current altimeter setting must be dialed into the altimeter.
The IFR approach is ILS runway 24 right, tower frequency 118.35, and ILS runway 23, tower frequency 118.7.	For aircraft approaching Toronto to land, the runways numbered 24 right and 23 are in use. The appropriate radio frequency for each runway should be used to communicate with the tower in preparation for landing. Runway numbers are based on the runway's orientation relative to a 360 degree compass (where 360 is north, 090 is east, 180 is south, and 270 is west). For example, an aircraft lined up for takeoff on a runway numbered 36 (shortened from 360) would be facing north. The other end of that same runway would be numbered 18 (shortened from 180), and an aircraft lined up for takeoff would be facing south. Refer to the photo.
Departure runways are 24 r	Aircraft preparing for departure should be advised that runways 24 right and 23 are active. Large airports often have parallel runways. In this situation the words *right* or *left* are used to distinguish between them. *ight and 23.*
Inform Toronto ATC on initial contact that you have received information CHARLIE.	When pilots make initial contact with air traffic control. they should inform them that they have information "Charlie."

Figure 6.7 *Runway Number*
In this picture, the runway is oriented towards 080 degrees. Therefore, the runway is aligned slightly (10 degrees) north of east.

VFR Controllers

The next group that a flight will interact with is the VFR controllers. These controllers work in airport control towers and provide approaching and departing aircraft with clearances and instructions to ensure adequate spacing from other aircraft (horizontal, lateral and vertical). Controllers also ensure separation between aircraft and vehicles operating on the maneuvering area of the airport. This is done using visual control methods sometimes supplemented by air and ground surveillance systems to monitor movements. VFR controller positions include clearance delivery (although this ATC position is only at major airports), ground control, and tower control.

Clearance Delivery

Before an airplane can begin taxiing for an IFR flight out of a major airport, the pilot must receive a clearance from a controller. Clearance controllers are located at the airport in the control tower. Upon first contact, pilots will inform the controller that they have the current ATIS information and let them know who they are, where they are, and what they want to do. The controller working clearance delivery will find the flight plan each pilot filed before flight. All IFR flights and some VFR flights require flight plans. The clearance controller will ensure that pilots have the latest ATIS information, verify flight plans to ensure that they are valid and in their system, the aircraft are on the right transponder frequency, and there are no conflicts with other aircraft. If a conflict is found, the controller may amend a flight plan, impose flow restrictions, or delay the takeoff clearance until the potential conflict has passed. At that point, the controller will issue a clearance to the pilot. Pilots will write down this important information which includes the radio frequency that they must tune in upon departure, transponder code, altitude, route, and departure procedure.

Ground Control

Once these steps have been completed, the aircraft is ready to begin taxiing. Before moving, the pilot must contact ground control. Ground controllers are responsible for surface movements on the airport, including taxiways, runway aprons, and inactive runways. All maneuvering areas on the airport are controlled by ground controllers, with the exception of the active runway and privately owned areas. All aircraft, vehicles, and persons require permission from ground control to operate in these areas. This permission is typically granted through a radio, which will be on board all airport vehicles (including aircraft, fuel trucks, emergency services, baggage carts, etc.).

The goal of ground controllers is to assist in the safe and orderly movement of vehicles and aircraft around the airport and to avoid runway incursions. A runway incursion is defined as "any occurrence at an airport involving the unauthorized or unplanned presence of an aircraft, vehicle, or person on the protected area of a surface designated for aircraft landings and departures" (Aviation Safety Letter 2002).

Ground controllers are located in the airport tower. The tower is located on the airport in a high location from which the majority of the airport is visible. Ground controllers often use binoculars to visually keep track of surface movements of aircraft and vehicles in order to avoid conflicts.

A leading-edge ground control technology in use by NAV CANADA is called Multistatic Dependent Surveillance (MDS). MDS makes use of several sensors that triangulate aircraft location based on their transponder signal. This allows controllers to have an enhanced display of aircraft ground position compared to traditional surface radar. This information is particularly important for controllers during poor-visibility conditions (during fog or snowstorms). Through communications with the ground controller, the aircraft will be led safely to hold short of the active runway in preparation for takeoff.

Tower Controller

The tower controller is responsible for movements on the active runway surface and flights in the control zone. The control zone is an area surrounding the airport, usually by seven nautical miles, and overlying the airport, usually by 3000 feet. The control zone constitutes the smallest and lowest level of the upside-down wedding cake.

The term *tower controller* is a misnomer because clearance and ground controllers are also located in the tower!

No aircraft is permitted to land or take off without permission from the tower controller. Once an aircraft is cleared for takeoff and has begun its en route leg, and has left the control zone, the tower controller will tell the pilot to contact departure.

IFR Controllers

The next ATC group that a flight will interact with is the IFR controllers. IFR controllers work in one of the seven Area Control Centres (ACCs) in Vancouver, Edmonton, Winnipeg, Toronto, Montreal, Moncton, and Gander. IFR controllers give instructions, information, and clearances to pilots during the en route part of their flight. They use surveillance systems, such as radar, and communications to track flights within their ACC's airspace.

The workspaces of IFR controllers are much different from those of VFR controllers. Rather than visually looking out of a control tower, IFR controllers are located in a windowless room at the ACC and track aircraft by looking at a radar screen or using flight data boards with FPSs.

Each terminal control area (TCA), described previously in this chapter as resembling an upside-down wedding cake, has an associated terminal control unit (TCU). Terminal (TCU) controllers manage the departure, arrival, and transitioning traffic for the TCA. TCUs are located within the seven ACCs. Each ACC is responsible for a volume of airspace called a flight information region (FIR). ACCs provide air traffic control, information services and alerting services for aircraft within their FIR.

Departure Control

After departing aircraft have been told to contact departure, they will tune their radio to a departure frequency and establish communications with a controller in a TCU. TCU controllers handle both approaches and departures within the terminal control area (TCA) associated with a major airport. Although a TCU controller is commonly referred to as "departure" or "approach," the same person may answer both calls (depending on the busyness of the area). The goal of TCU controllers is to safely separate and sequence traffic. This is accomplished exclusively with information on a radar screen, information technology, and voice communication. These controllers have no way of visually seeing the aircraft in the sky. During departure, once the TCU controller says that

Copyright NAV CANADA 2009.

Figure 6.8 *Toronto Air Traffic Control Tower*

TENERIFE AIRPORT DISASTER

By Joshua Norman

On 27 March 27 1977 at 1230 hrs. local time, a bomb exploded outside the Gran Canaria Airport located in Spain's Canary Islands. This explosion forced the diversion of Pan Am Flight 1736 and KLM Flight 4805, both Boeing 747 aircraft, to Tenerife Airport. Tenerife Airport, located forty miles west of Gran Canaria, consists of one runway with a parallel taxiway. After approximately three hours, Gran Canaria was reopened to traffic, and both the Pan Am flight and the KLM flight prepared to depart Tenerife Airport for their final destinations. Due to increased congestion caused by the multiple diversions from Gran Canaria, both aircraft were required to backtrack to their positions on the runway prior to takeoff.

While the Pan Am B747 was backtracking on the runway to an exit taxiway, the KLM B747 captain began applying takeoff thrust, but he stopped when questioned by the first officer as to whether or not they had received a takeoff clearance. The captain, who was one of the most experienced pilots flying for the airline at the time, throttled down and told the first officer to request a clearance. When the first officer contacted ATC, he was given a departure clearance but was not yet cleared to take off. All air crew aboard the KLM flight mistakenly believed the departure clearance constituted a clearance for takeoff. This confusion was attributed primarily to the use of nonstandard language by the air traffic controller and KLM first officer. A few seconds later, after beginning their takeoff roll, KLM heard a message from the Pan Am aircraft to ATC stating that they would report when clear of the runway. The KLM flight engineer asked the captain if the Pan Am aircraft was not clear of the runway and the captain emphatically said that it was. It is believed that the flight engineer did not further question the situation due to the captain's perceived authority in the cockpit and status in the company.

Due to low visibility caused by fog, the KLM air crew was not able to see that the Pan Am B747 was still taxiing on the runway when they started their takeoff roll. Approximately 9.5 seconds prior to impact, the Pan Am air crew saw the KLM barrelling down the runway towards them. At the same time, the KLM aircraft pitched up aggressively in an attempt to climb over the Pan Am aircraft. At 1706 hours, the KLM aircraft's landing gear and right wing contacted the Pan Am aircraft, ripping off its roof and upper deck. The right wing and landing gear were sheared off the KLM aircraft and it crashed 500 metres passed the initial point of impact. All 248 on board the KLM aircraft perished, and only sixty-one of the 326 aboard the Pan Am B747 survived. In total, 583 people were killed in the accident, which makes the Tenerife disaster the most deadly aviation accident in history.

The findings of Spain's Subsecretaria de Aviacion Civil accident investigation placed a heavy emphasis on the role of human factors in this accident. They also placed sole responsibility of the accident on the KLM captain for taking off without clearance and not aborting takeoff when there was a possibility that the Pan Am aircraft was still on the runway. The Netherlands Aviation Safety Board, on the other hand, refuted the claims and placed more blame on generally accepted procedures that were used in the aviation community. Both parties agreed that the nonstandard language used by all air crew and Tenerife ATC contributed to the accident. Low visibility and the stress placed on Tenerife ATC by unusual traffic congestion also played a role.

As a result of this accident, sweeping changes were made throughout the aviation industry. These included the standardization of language used in departure and takeoff clearances and a stronger emphasis on research in the realm of human factors. This runway incursion was a deadly example of the importance of precise ground control and standard radio procedures to safe flight operations.

Copyright NAV CANADA 2009.

Figure 6.9 *Vancouver Area Control Centre (ACC)*

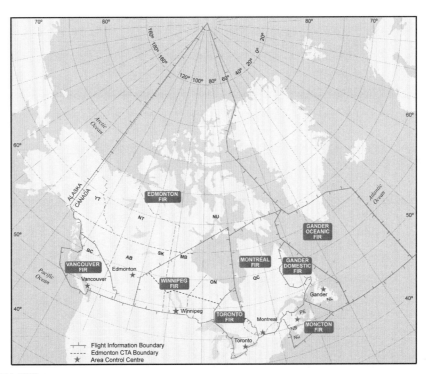

Copyright NAV CANADA 2009.

Figure 6.10 *Map of Flight Information Regions (FIRs)*

the aircraft is "radar identified," it is now legally the controller's responsibility to ensure safe separation from other IFR traffic. Once the flight has cleared the airport's terminal control area (TCA), departure control will ask the pilot to contact centre.

Centre Control

The last group of controllers that a flight will communicate with is called *centre*, These controllers work within one of the seven area control centres (ACCs) spread out across Canada. By far, centre controllers make up the majority of controllers at NAV CANADA.

Centre controllers will handle a flight until it passes out of their FIR, at which point they will hand the traffic over to an adjacent ACC or transition it to the TCU, located within an ACC, which is responsible for traffic at the arrival airport. After passing through the centre controller's region, the pilot would then speak to approach, tower, and eventually the ground controller until they come to a stop at the gate.

En route Navigation

In many cases aircraft operating under instrument flight rules, which includes most commercial air transportation, do not simply fly a straight line from their departure airport to their destination. Rather, they are required to follow predetermined standardized routing. En route, this system is termed *airways*. These airways can be visualized as highways in the sky. Each airway has a name and special rules for entering or leaving. When airways cross over each other, invisible intersections are formed. These intersections also have names. While utilizing these intersections, you will be in contact with ATC and given directions on how to proceed.

Airways are formed based on linkages between ground-based navigational aids. Examples of these navigation aids include VHF omnidirectional range (VOR), distance measuring equipment (DME), and non-directional beacons (NDBs). VOR/DMEs provide the most common navigational aid used in the formation of airways. An airway based off of a VOR signal is called a *Victor airway*. Victor airways begin at 1000 feet above ground level and go up to, but not including, 18,000 feet above sea level. These are considered low-level airways. High-level airways, sometimes called *jet airways,* begin at 18,000 feet above sea level and go up to FL 450.

Another method of en route navigation that is increasingly utilized is Area Navigation (RNAV). RNAV allows aircraft to choose any desired course within the network of ground-based navigational beacons rather than directly from one beacon to the next. This method allows for significantly shorter flight distances between airports, as, once at altitude, an aircraft can fly directly to its destination, avoiding the zigzag of airways. RNAV waypoints that are strung together to form a highway in the sky are called *T-routes*.

In some areas, radar coverage is not available, such as in sparsely populated areas or over large bodies of water. In this situation, ATC utilizes procedural control. Procedural control utilizes information regarding the time and the aircraft's airspeed and altitude while flying on predetermined paths. Based on this information, ATC spaces aircraft approximately ten to fifteen minutes flying time apart. This is significantly more separation than is required while radar identified, and slows down the ANS.

In Canada, Hudson Bay encompasses a 250,000 square kilometre area where radar coverage is not available. In January of 2009, NAV CANADA brought surveillance coverage to this area through ADS-B technology. ADS-B is the single greatest change to air navigation surveillance since the implementation of radar. ADS-B stands for an Automated Dependent Surveillance-Broadcast system. The system is "automatic" because it sends position reports automatically without requiring any action from the pilot. It is "dependent" because it requires aircraft onboard equipment to provide "surveillance" information which is "broadcast" to ATC.

ADS-B is not radar, but it supports the same efficient separation standards as radar surveillance and is more accurate and affordable. For an aircraft to operate under ADS-B surveillance, onboard navigation equipment is required. This equipment works with satellites, ground stations, and communication systems to give ATC position information. Avoiding the greater separation required with procedural control, aircraft utilizing ADS-B over Hudson Bay are expected to save fuel costs, flight times, and the resulting fuel emissions. As approximately

35,000 flights utilize this airspace each year, these benefits are expected to equate to eighteen-million litres of fuel and reductions in CO_2 emissions of 50,000 tonnes per year once all aircraft flying over Hudson Bay are equipped for ADS-B navigation.

Military pilots utilize a similar navigational procedure, although they often utilize civilian airways as well. The military has a separate navigational system, including an independent network of airways. The military system is called *tactical air navigation* (TACAN). TACAN allows for more accurate navigation than what is available in civil aviation. Basically, TACAN can be considered an enhanced VOR. TACAN stations are typically located in the vicinity of military airports. However, the distance measuring system in TACAN is the same as in the civilian world. Therefore, to minimize the number and maintenance of ground-based navigational facilities, sometimes TACAN facilities are co-located with VOR stations. These are known as VORTACs.

Safety Authority

After gaining an understanding of how air navigation is accomplished, it may raise several questions, such as: Are pilots legally required to follow the instructions of air traffic controllers? What happens if a pilot recognizes that an ATC instruction is a mistake or it may put them at risk for a collision? Are they legally allowed to disregard ATC instructions in the interest of safety?

Being human, controllers will make mistakes. To answer these questions, CARS 602.31 explains the "compliance with air traffic control instructions and clearances" as the following:

(1) SUBJECT TO SUBSECTION (3), THE PILOT-IN-COMMAND OF AN AIRCRAFT SHALL

(A) COMPLY WITH AND ACKNOWLEDGE, TO THE APPROPRIATE AIR TRAFFIC CONTROL UNIT, ALL OF THE AIR TRAFFIC CONTROL INSTRUCTIONS DIRECTED TO AND RECEIVED BY THE PILOT-IN-COMMAND; AND

(B) COMPLY WITH ALL OF THE AIR TRAFFIC CONTROL CLEARANCES RECEIVED AND ACCEPTED BY THE PILOT-IN-COMMAND AND

(I) SUBJECT TO SUBSECTION (2), IN THE CASE OF AN IFR FLIGHT, READ BACK TO THE APPROPRIATE AIR TRAFFIC CONTROL UNIT THE TEXT OF ANY AIR TRAFFIC CONTROL CLEARANCE RECEIVED, AND

(II) IN THE CASE OF A VFR FLIGHT, READ BACK TO THE APPROPRIATE AIR TRAFFIC CONTROL UNIT THE TEXT OF ANY AIR TRAFFIC CONTROL CLEARANCE RECEIVED, WHEN SO REQUESTED BY THE AIR TRAFFIC CONTROL UNIT.

(2) EXCEPT IF REQUESTED TO DO SO BY AN AIR TRAFFIC CONTROL UNIT, THE PILOT-IN-COMMAND OF AN IFR AIRCRAFT IS NOT REQUIRED TO READ BACK THE TEXT OF AN AIR TRAFFIC CONTROL CLEARANCE PURSUANT TO PARAGRAPH (1)(B)(I) WHERE

(A) THE AIR TRAFFIC CONTROL CLEARANCE IS RECEIVED ON THE GROUND BY THE PILOT-IN-COMMAND BEFORE DEPARTING FROM A CONTROLLED AERODROME IN RESPECT OF WHICH A STANDARD INSTRUMENT DEPARTURE PROCEDURE IS SPECIFIED IN THE *CANADA AIR PILOT;* OR

(B) THE RECEIPT OF THE AIR TRAFFIC CONTROL CLEARANCE IS ACKNOWLEDGED BY THE PILOT-IN-COMMAND BY ELECTRONIC MEANS.

(3) THE PILOT-IN-COMMAND OF AN AIRCRAFT MAY DEVIATE FROM AN AIR TRAFFIC CONTROL CLEARANCE OR AN AIR TRAFFIC CONTROL INSTRUCTION TO THE EXTENT NECESSARY TO CARRY OUT A COLLISION AVOIDANCE MANOEUVRE, WHERE THE MANOEUVRE IS CARRIED OUT

    (A) IN ACCORDANCE WITH A RESOLUTION ADVISORY GENERATED BY AN AIRBORNE COLLISION AVOIDANCE SYSTEM (ACAS) OR A TRAFFIC ALERT AND COLLISION AVOIDANCE SYSTEM (TCAS); OR

    (B) IN RESPONSE TO A WARNING FROM A GROUND PROXIMITY WARNING SYSTEM (GPWS) ON BOARD THE AIRCRAFT.

(4) THE PILOT-IN-COMMAND OF AN AIRCRAFT SHALL

    (A) AS SOON AS POSSIBLE AFTER INITIATING THE COLLISION AVOIDANCE MANOEUVRE REFERRED TO IN SUBSECTION (3), INFORM THE APPROPRIATE AIR TRAFFIC CONTROL UNIT OF THE DEVIATION; AND

    (B) IMMEDIATELY AFTER COMPLETING THE COLLISION AVOIDANCE MANOEUVRE REFERRED TO IN SUBSECTION (3), COMPLY WITH THE LAST AIR TRAFFIC CONTROL CLEARANCE RECEIVED AND ACCEPTED BY, OR THE LAST AIR TRAFFIC CONTROL INSTRUCTION RECEIVED AND ACKNOWLEDGED BY, THE PILOT-IN-COMMAND.

Ultimately, flight safety is the responsibility of the pilot. This is very important to understand, because it is the pilot, not necessarily the air traffic controller, who will be blamed if a mid-air collision or controlled flight into terrain (CFIT) occurs. Therefore, if a pilot notices an unsafe situation, he or she is required to deviate from ATC instructions to complete an avoidance maneuver and then notify ATC as soon as possible once the flight is stabilized.

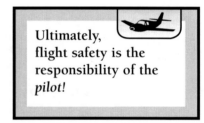

Ultimately, flight safety is the responsibility of the *pilot!*

Career Paths

IFR Air Traffic Controller:

These controllers are based in one of the seven area control centres (ACCs) across Canada. IFR ATCs track flights within their sector, provide en route instructions to pilots, and provide airport clearances.

VFR Air Traffic Controller:

These controllers are based in one of the forty-one ATC airport towers across Canada. VFR ATCs visually watch every approach, departure, and taxiway movement and give clearances for landing, take-off, and taxi. Once an aircraft has taken off, the flight is handed over to an IFR ATC.

Flight Service Specialist:

These individuals work in flight information centres (FIC) and flight service stations (FSS) across Canada. They provide weather briefings, file flight plans, communicate safety updates, help pilots safely operate out of uncontrolled airports, provide weather reports to Environment Canada, and trigger search-and-rescue for missing or overdue aircraft.

Salary:

Average ATC salary—$75,000/year

Trainee—$30,000–33,000/year

IFR and VFR Air Traffic Controllers—$57,000–150,000/year

FSS—$42,000–86,000/year

Selection Requirements:

Applicants must be Canadian citizens or permanent residents who have a minimum of a high school education. Applicants must agree to a medical examination and security check. Otherwise, no specific experience is required, as all selected applicants will be fully trained by NAV CANADA. Selection is based on a series of tests that assess abilities, aptitudes, and character traits. Selection procedures are designed to identify people with the best potential to succeed as a controller or flight service specialist. Applicants must pay a small fee to attempt the selection tests.

Training:

All training is delivered at an area control centre (ACC). However the duration of training varies based on position. After completing training, students may be posted anywhere in Canada, based on the company's operational requirements.

IFR ATC: Training lasts seven to fourteen months and is followed by paid on-the-job training of six to twelve months. Tuition is $3,500.

VFR ATC: Training lasts four to six months and is followed by paid on-the-job training of four to twelve months at an air traffic control tower. Tuition is $2,500.

FSS: Training lasts up to six months and is followed by up to six months of on-the-job training. Tuition is $1,000.

continues

Licensing:

Air traffic controllers are required to maintain an ATC license. These licenses are issued by Transport Canada. To be issued a license, a controller is required to:

- Obtain a valid medical certificate
- Prove citizenship
- Pass a knowledge exam
- Have sufficient experience

Lifestyle:

Controllers begin their position with an enviable salary and twenty-six days of vacation each year. Throughout their career, controllers work seventeen out of every twenty-eight days.

Conclusion

Canada's air navigation system (ANS) is the second largest in the world. Air navigation is a complex system that involves three main components: infrastructure, people, and equipment. The infrastructure of Canada's ANS is established and maintained by several groups, including Transport Canada (who develops and enforces regulations) and airport authorities. The people and equipment of Canada's ANS are managed by NAV CANADA.

NAV CANADA, being a private entity, makes the ANS in Canada unique in the world. Since it accepted responsibility for Canada's ANS, NAV CANADA has continually strived to increase safety, modernize equipment, and maximize efficiency.

PRACTICE ACTIVITY

Practice decoding this sample ATIS from the Vancouver International Airport.

"*Vancouver International Airport information WHISKEY, Vancouver weather at 1842 Zulu, wind 240 at 12, visibility 8, ceiling 4500 overcast, temperature 11, dewpoint 10, altimeter 2978. The IFR approach is ILS runway 26 left and runway 26 right. Simultaneous parallel ILS approaches in use. Departure runway is 26 left. Inform Vancouver ATC on initial contact that you have received information WHISKEY.*"

ATIS Text	Description
Vancouver International Airport information WHISKEY	
Vancouver weather at 1842 Zulu	
Wind 240 at 12	
Visibility 8	
Ceiling 4500 overcast	
Temperature 11, dewpoint 10	
Altimeter 2978	
The IFR approach is ILS runway 26 left and runway 26 right. Simultaneous parallel ILS approaches in use.	
Departure runway is 26 left	
Inform Vancouver ATC on initial contact that you have received information WHISKEY	

CHAPTER 7

HUMAN FACTORS

OBJECTIVES

After you have read this chapter, you will:

- Understand and be able to define the science of human factors

- Be able to recite the IMSAFE checklist and describe each component

- Know the five approaches that human factors experts use to solve problems

- Be familiar with Crew Resource Management (CRM) and Single Pilot Resource Management (SRM)

Introduction

Human factors refers to the study of all aspects of human beings (physical, psychological, and emotional) with the purpose of maximizing job performance. Academically, the science of human factors draws from several disciplines, including ergonomics, physiology, psychology, engineering, management and education. Compared to these other disciplines, the field of human factors is relatively new. It was born out of necessity during World War II. The demand for aviators to fight in the war grew so quickly that people started researching how to train pilots in less time and design aircraft that were easier to fly.

EXAMPLE OF BAD HUMAN FACTORS

At the end of World War II, pilot Dan Bowman admitted to an incident that occurred during the war. He was on call, suited and ready to fly when the air raid warnings went off, signalling that Japanese fighters were inbound to attack the base. He started running to a P-47 Thunderbolt on the tarmac and was the last pilot to reach a plane. The plane he climbed into was new to the base, as it had arrived only two days before. When he looked around the cockpit, he noticed that something was very wrong. None of the flight instruments were in the right spot, the entire cockpit had been redesigned, and Dan could not figure out how to start the plane. After a few minutes, he finally figured out how to start the engine—just as a Mitsubishi Zeke fighter bomber was approaching to strike. Dan realized he couldn't take off in this plane, so he taxied the aircraft out of the attack zone of the approaching bomber. For the next twenty minutes of raids, Dan continued to taxi the Thunderbolt around the base, across the runway, and along the taxiways until the raid was finally over. Since this time, human-factors scientists have gained an understanding of the ramifications of a complete redesign of a cockpit in the middle of a war (Casey 1998).

Republic P-47B "Thunderbolt"
USAF Museum Photo Archives
USAF Museum Archives.
Figure 7.1

When learning about human factors, it is easy to become overwhelmed, as it is such a broad topic. To simplify human factors, the IMSAFE checklist is commonly used in aviation training. The IMSAFE checklist is an acronym that quickly outlines human factors topics that a person should consider before piloting an aircraft. The acronym stands for illness, medication, stress, alcohol, fatigue, and eating.

Illness

Before pilots fly, they should ask: Am I sick? Do I have symptoms of being ill? Similar to many industries, such as firefighting or the police force, there has historically been a "macho" culture in aviation where pilots are encouraged to "tough-it-out" through minor and moderate illnesses. Ask any new commercial pilot what the chief pilot would say if he or she called in sick because of a slight cough, and you could probably expect some laughter to be included in the response. However, an untreated cold will reduce reaction time by 57 percent and impair performance to the same extent as a couple of beers (Reinhart 1996). Attitudes toward illness in the aviation industry are slowly changing as more is learned about how even minor sickness can impair performance. Although some aspects of health, such as genetics, are outside our

I	=	*Illness*
M	=	*Medication*
S	=	*Stress*
A	=	*Alcohol*
F	=	*Fatigue*
E	=	*Eating*

control, there are things a pilot can do to maintain good health. These include living a healthy lifestyle, maintaining a healthy diet, and exercising regularly.

Another aspect to consider is mental illness. Pilots are an unusual group in this regard. For example, when researchers administer psychological questionnaires (with questions such as "sometimes I feel stress" or "sometimes I become irritable"), most people have a moderate result, indicating that they occasionally experience these emotions. Pilots tend to score around zero. They indicate that they never experience any of these emotions. This can be interpreted as an influence of the occupational culture of aviation that encourages pilots to "tough-it-out." However, in reality pilots are just as susceptible to depression, anxiety, stress, and phobias as other human beings. In some regards, they are at greater risk due to demanding work, travel, and training schedules. Overall, it is important that mental illnesses be taken seriously, as they put flight safety at risk. Most airlines provide a twenty-four-hour toll-free number to their air crew to reach psychiatric help, just in case help is needed. Pilots are encouraged to call this line if they encounter any personal or professional situations that may impact their psychological state.

MENTAL ILLNESS IN FLIGHT

On 28 January 2008, an Air Canada flight from Toronto to London's Heathrow International Airport had to be diverted to Shannon, Ireland. The reason for the diversion was that the first officer became ill during the flight. It was believed that he suffered a nervous breakdown en route, as he began swearing loudly and asking for God. He had to be restrained by the cabin crew and a passenger who was an off-duty member of the Canadian military. After landing, the first officer was taken to an Irish psychiatric hospital (Nguyen 2008). Thankfully, the safety of the flight was not compromised. However, the cost to the airline for a diversion is staggering when extra fuel, landing fees, passenger hotels and meals, rebooking passenger connections, and new crew expenses are considered.

Medication

Before flight, pilots should consider: Have I taken any prescription or over-the-counter medications? Have I discussed these medications with an aviation medical examiner (AME)? Pilots are required to pass medical evaluations on a recurring basis in order to maintain their licenses. However, due to the "tough-it-out" attitude discussed previously, some aviators choose to self-diagnose a minor illness and self-medicate with over-the-counter (OTC) cough, cold, or pain medication. Typically, this is choice is made out of fear of losing their aviation medical certificate had they visited an AME. In recent years, an investigation found hundreds of fatal accidents in which pilots had withheld potentially disqualifying medical conditions and chose instead to medicate themselves (roughly 10 percent of all accidents over a ten-year period) (Committee on Transportation and Infrastructure 2007). There are many problems associated with the decision to withhold information from AMEs and self-diagnose and self-medicate. For example, over-the-counter medication can significantly impact a pilot's cognitive and psychomotor skills. Some of these problems can include decreased visual and auditory perception, impaired decision making and judgment, and reduced hands-on flying skills (Reinhart 1996). All OTC medications are tested and determined to be safe for the general public. However, this does not extend to the aviator's environment, as that is beyond the typical safety evaluation. As a general rule of thumb, pilots should speak with an AME about any medication they plan to use during flight.

DANGER OF UNREPORTED MEDICATION USE

On 26 November 1999, Itzhak Jacoby was flying his Beechcraft aircraft home to Washington, D.C. Itzhak Jacoby was accompanied by his wife Gail, and Atira, their daughter. While en route, the aircraft crashed into a residential neighbourhood in New Jersey. The crash fatally injured all three occupants, injured twenty-five people on the ground, and caused $1.2 million in property damage. The pilot's autopsy identified a drug in his system called *Fiorinal,* which contains barbiturates and is used to treat acute migraines. Symptoms from the use of this drug include reduced coordination, slower thinking, decreased memory, and poor decision making. During Itzhak Jacoby's last medical evaluation, one month earlier, he stated he was not taking any prescription or OTC medication and did not suffer from headaches. His personal medical history revealed that he had been diagnosed with severe migraine headaches and prescribed over 6000 caplets of Fiorinal. The investigation determined the cause of the accident was Itzhak Jacoby's medical condition and medication (Committee on Transportation and Infrastructure 2007).

Stress

Prior to flying, pilots should review: Am I being influenced by any stress? Job, family, financial, or health related? Am I wearing appropriate clothing for the weather to avoid heat and cold stress? Stresses can be either environmental or psychological in nature. Environmental stresses include temperature, noise, and lighting level. Psychological stresses include fear, anxiety, and peer pressure. Stress will impact your ability to pilot an aircraft safely. Stress can impact your ability to concentrate on tasks because it is a distraction. For example, although you should be concentrating on the flight, you may find your mind wandering to an argument you had with your spouse that morning. There are other times that stress has a physiological impact, such as sweating and/or stomach pains in anticipation of an interview or speech. In any case, it is important to be aware of how you respond to stressful situations. As an individual, you can expect to react to stress in a similar way each time. Understanding your reaction is important, as it allows you to recognize the amount of stress you are experiencing. Over time, this recognition will allow you to develop coping strategies. These strategies may include distraction, breathing techniques, exercise, visualization, or exercise.

Alcohol

Before planning a flight, pilots should think: Has it been eight hours since I last consumed alcohol? Has it been twenty-four hours? The Canadian Aviation Regulations (CARs) section 602.03 states that:

> NO PERSON SHALL ACT AS A CREW MEMBER OF AN AIRCRAFT (A) WITHIN EIGHT HOURS AFTER CONSUMING AN ALCOHOLIC BEVERAGE; (B) WHILE UNDER THE INFLUENCE OF ALCOHOL; OR (C) WHILE USING ANY DRUG THAT IMPAIRS THE PERSON'S FACULTIES TO THE EXTENT THAT THE SAFETY OF THE AIRCRAFT OR OF PERSONS ON BOARD THE AIRCRAFT IS ENDANGERED IN ANY WAY (TRANSPORT CANADA 2008 ¶3).

Section (a) of this regulation has spawned the aviation expression "8 hours from bottle to throttle" which means that the last alcoholic drink a pilot consumes must be finished at least eight hours before resuming flight activities. Section (b) expands upon the first point of the regulation to include anytime a person is under the influence of alcohol. The confusion with this is that, depending upon your level of intoxication, you may still be under the influence of alcohol 8 hours after your last drink. In fact, even when your blood alcohol level is 0, impairments from hangovers and other negative physical effects remain. These effects are still classified as the "influence of alcohol" and operating an aircraft hung-over would be in violation of the regulation. Although this regulation clearly restricts alcohol consumption, it is important to note that section (c) expands the limitation to any drug that impairs a person's faculties. By this definition, most OTC medications will fall under this category. This is further justification to check with an AME before using medication during piloting activities.

EXAMPLE

Remember that all alcohol drinks do not contain the same amount of alcohol. In fact, 12 oz of beer = 1 1/2 oz of liquor = 5 oz of wine. The more alcohol in your system, the longer it will take to metabolize out of your body and the longer you will be under the influence.

© Scott Pehrson Shutterstock.

Figure 7.2

Fatigue

Before the flight, pilots should ask: Am I adequately rested? Do I expect to be adequately rested at my estimated time of arrival (ETA)? Fatigue is a significant safety issue in aviation and is an unavoidable condition for pilots. This is due to the nature of the occupation, which includes sleep loss, high workload, monotonous activities, jetlag, stress, and long duty weeks. Fatigue is not something that pilots can "tough-out" any more than a passenger on a long-haul flight could. It is a natural limitation associated with being a human being. The detrimental effects of fatigue include reduced attention, decreased concentration, increased errors, greater risk acceptance, and longer response times (Bourgeois-Bougrine, Carbon, Gounelle, Mollard, and Coblentz 2003; Caldwell, Hall, and Erickson 2002).

Research explains fatigue as the result of sleep loss and disrupted circadian rhythms (Mann 1999). It is important for a professional pilot to understand that losing sleep has serious implications to the safety of flight. Therefore, it is every pilot's responsibility to take safety seriously and ensure a full night's rest before flight. Circadian rhythms are the natural twenty-four-hour biological clock that affects every human being. This clock impacts our appetite, mood, and energy levels and is directly linked to the daylight/night-time cycle where we live. Pilots may experience disruptions to their circadian rhythms, commonly in the form of jetlag, during long-haul missions. Symptoms of jet lag include loss of appetite, depressed mood, poor coordination, insomnia, reduced alertness, and impaired thinking skills (Spitzer et al. 2008). To combat jetlag, there are a few strategies to adopt. If it is a short stopover, you should attempt to maintain your home routines (even if it means eating breakfast in the middle of the night). If you are staying at your destination for a few days, the following are recommended:

- If you need to be at 100 percent performance when you arrive at your destination, you can expose yourself to bright light during your destination's daylight hours before leaving home.

- Upon arrival, adjust to your destination's local time as quickly as possible. If you arrive early in the day, wait until the local bedtime to sleep.

- Begin eating meals at local mealtimes.

- Be outside and active during your destination's daylight hours.

CASE

ASLEEP AT THE WHEEL

In March of 2004 a dangerous incident occurred in the skies over Colorado. The captain and first officer of an Airbus 319 headed to the Denver International Airport both fell asleep. The captain anonymously reported this information to the Aviation Safety Reporting System (ASRS) (more information about ASRS is provided in Chapter 8). ASRS allows pilots to submit reports of dangerous incidents without risk of prosecution. The pilot reported that he flew a red-eye (overnight) flight, after two previous red-eye flights, and, after only a one-hour break, immediately started the seven-hour flight back to Denver. In the last forty-five minutes of the flight, he fell asleep, and so did the first officer. He missed all of the calls from ATC, crossing a navigational intersection 16,000 feet too high and 350 nautical miles per hour too fast. The captain eventually woke up, although he wasn't sure what woke him, and heard frantic calls from ATC. He then woke up the first officer and they were able to land the aircraft without further incident (ASRS 2004). As this report is anonymous, it is unknown what airline this pilot was working for. However, as this incident was reported to the airline, it is expected that the routing and pairings of trips was changed to avoid further problems with fatigue.

Eating

Prior to flying, a pilot should consider: Have I eaten and drunk enough to last me until the estimated time of arrival (ETA)? This is a factor that is easy to overlook. With all of the pre-flight preparations, it can seem inconsequential to throw a bottle of water or a snack into your flight bag. However, after flying for several hours without ingesting any food for energy, a pilot is prone to a hypoglycemic state. Hypoglycemia is the result of inadequate blood sugar in the body. This could result in several symptoms, including shakiness and generally feeling ill. The immediate remedy for this situation is to eat something high in sugar/carbohydrates. However, the best treatment for this situation is prevention. Although pilots travel frequently, they need to maintain good eating habits. In general, try to avoid foods with low nutritional value and maintain a balance of proteins, carbohydrates, and fats at each meal. Hydration is also critically important, particularly in hot environments. However, in a general aviation aircraft, too many fluids can also cause a problem, as no restroom is readily accessible. The best practice is to consume fluids slowly over the duration of the flight.

PROFESSIONAL PROFILE

Title:

Aviation Human Factors Expert

Educational and Work Experience:

Human factors experts typically require advanced education, including a Master's degree or a Ph.D. These experts focus in one of the following academic disciplines: psychology, ergonomics, engineering, education, or human factors.

These experts generally work for aircraft or avionic manufacturers to engineer human friendly systems, in research organizations to evaluate where human errors are occurring and to recommend resolutions, or in universities to teach students about human factors and perform research part-time.

Due to the requirement for advanced education, people are typically twenty-five to thirty years old before beginning work as a human factors expert. Very often a person will begin work as a junior human factors specialist and work his or her way up.

Type of Work:

This work is very stable, provides regular hours and sometimes the ability to work from home. Human factors experts who specialize in research are typically given the freedom to investigate what they are interested in. Overall, this is a job that experts enjoy because (unlike other academic disciplines) their work applies directly within the real world. This means that findings from research are often used to develop safety interventions in airlines or the military.

Salary:

$80,000–100,000/year

Human Factors Approach to Problems

At its core, there are five approaches a human factors expert uses to solve problems. These approaches are equipment, task, environment, selection, and training (Wickens, Lee, Gordon, and Liu 2003).

Equipment

This approach involves modifying the physical equipment that a person works with. For a pilot, this equipment includes large items like the aircraft down to a very small item like a pilot's pen. Any piece of equipment that is potentially faulty, distracting, difficult to use, cumbersome, or confusing should be redesigned, as it draws attention away from the pilot's primary task of flying the aircraft. Examples of equipment that have been redesigned in aviation include:

- Replacing old steam-engine avionic displays with easy-to-read glass cockpit displays
- Placing a safety guard over a helicopter's load release button so that it cannot be accidentally triggered
- Using a pitot static cover that is large and colourful so that it is not easily missed

EXAMPLE

This is an example of how avionics have been redesigned. The first picture shows the traditional "six pack" steam-engine gauge avionic configuration. The second picture shows a glass cockpit where all of the information from the "six pack, and much more," is integrated into computerized displays. Glass cockpit displays allow pilots to gather and understand the flight parameters quickly and easily.

Deborah Aronds Shutterstock.

Figure 7.3

© Peter R. Foster IDMA Shutterstock.

Figure 7.4

Task

Modifying the task involves changing what the operator does, rather than the equipment they work with. In aviation, this is most commonly in the form of a company's standard operating procedures (SOPs). SOPs are developed by the management or Chief Flight Instructor in an aviation company to standardize the way their pilots react in certain situations. SOPs take over where Transport Canada regulations end, meaning that even if you comply with regulations, you might not be operating in the safest possible manner. SOPs take the guess work out of potentially complicated situations for individual pilots and allow everyone in an aviation company to act in a predictable and safe manner. From a human factors perspective, a benefit of SOPs is that, if an unsafe situation is identified (such as pilots continually coming in for landing too quickly), it is feasible to change SOPs so that everyone in the company can avoid the risk.

Environment

This approach involves changing the work environment. There are two segments to this approach: organizational and physical.

Organizational

The *organizational* environment has a powerful influence on human performance. This organizational environment refers to the environment created by the influence of managerial styles, social norms within the organization, incentives offered to employees, and organizational morale. An example of how this can impact safety is an airline company offering huge incentives to employees for on-time arrivals. Although employees are motivated to work more quickly and avoid delays, speed often decreases accuracy. This organizational environment would not be conducive to safe operations.

Physical

Changing the *physical* environment may include modifying the following characteristics:

- Lighting: Lighting levels need to be evaluated, as too much can cause glare and reduce productivity and too little can make gauges hard-to-read. Modifications include adding or removing lights.

- Noise: In aviation it is very difficult to avoid noisy situations, due to the nature of the airport. However, it is important to understand that there are risks associated with working in noisy environments. These risks include irritation, masking (when a noise covers up important communication), and health hazards. Noise in the work environment can be so irritating to workers that they become distracted from their task. A good example of this is a dripping faucet. Although not a loud noise, a drip can cause such annoyance that it becomes difficult to concentrate on tasks. Noise can also make it difficult to hear important communication in the workplace, which is called masking. For example, it is possible for the engine noise to mask important Air Traffic Control (ATC) instructions. Lastly, noise can impact one's health, as prolonged exposure to loud noises can shift a person's hearing threshold temporarily, meaning that the quietest sound the person could hear would be much louder than normal. If your career requires you to be exposed to loud noises over several years, it is possible that your hearing threshold would be permanently shifted. This is known as occupational deafness. In these noisy environments, design improvements include wearing ear protection and aviator headsets.

- Temperature: If the temperature of a workplace is uncomfortably hot or cold, it will impact the worker's performance. Adding individual temperature controls, suggesting appropriate clothing, or providing fans can help reduce the problem. Remember that, in order to be comfortable, room temperature needs to be around five degrees lower in winter months, as workers wear more clothing.

- Vibration: There are two types of vibration that workers may experience in aviation—high frequency and low frequency. High frequency vibration is what you may experience in a helicopter. This vibration may result in difficulty reading displays and flying with precision. Redesign suggestions include enlarging avionic displays or using seat cushions that absorb some vibration. Low frequency vibration is what you may experience during a flight if you encounter turbulence or if the aircraft begins rocking back and forth. The main risk associated with low frequency vibration is motion sickness. Motion sickness occurs any time there is a decoupling between your visual and balance systems. To fight motion sickness, look out the window so that what your eyes see matches the motion inputs sensed by your balance system. If possible, pilot the aircraft yourself. Alternatively, eating some ginger helps some people overcome their symptoms.

Selection

Selection refers to choosing people with the best physical and mental characteristics to do the job. The military is a leader in the area of pilot selection. It costs approximately $1 million to train a military aviator, so they invest a lot of resources into selecting candidates with the greatest chance of successfully completing training. This process begins by thoroughly analyzing the required knowledge, skills, and abilities of the job and developing selection tests or measures. These tests may include personality questionnaires, intelligence measures, physical fitness assessments, interviews, and flight simulator evaluations. The military is able to adopt this approach because there is a very large pool of candidates and the cost of failure is very high. Other aviation companies, such as airlines or flight schools, hire pilots who are already licensed and have many hours flight experience. In this situation, the cost of making a poor hiring decision is dramatically less, as they have not invested in training the individual from the ground up. These organizations tend to limit selection to an interview, which is a more subjective and less effective method of selection.

Training

Last, human factors specialists will design training interventions to solve recurring problems. Training is meant to convey knowledge or skills so that pilots are able to respond to situations in the most effective manner. Ongoing recurrent training is an important part of a pilot's career, whether as a brand new instructor pilot or a seasoned Boeing 747 captain. Aviation companies invest an enormous amount of resources into their training departments.

CRM is defined as "the effective use of all resources to achieve safe and efficient flight operations."

When seconds count . . .

Annual CRM Training pays off.

Transportation Safety Board of Canada.

Figure 7.5 *Annual CRM training prepares you for the unexpected.*

Crew Resource Management (CRM)

In the 1970s, a number of high-profile accidents occurred. After investigation, it was determined that pilot error was the primary causal factor of these accidents. In response, the National Aeronautics and Space Administration held a workshop in 1979 and invited representatives from the air transportation industry. The workshop was titled "Resource Management on the Flightdeck" (Helmreich, Merritt, and Wilhelm 1999). The purpose of the workshop was to investigate what was causing these human error accidents. The research presented at the workshop identified several aspects of human behaviour which contributed to these accidents, including decision making, leadership, and communication between the captain and first officer. Based on this information, Crew Resource Management (CRM) was developed as a training program to teach pilot teams how to effectively use these nontechnical human factors skills. The first CRM courses were developed and utilized by United Airlines in North America and KLM in Europe. CRM is defined as "the effective use of all resources (hardware, software, and liveware) to achieve safe and efficient flight operations" (Jensen 1995 116). Within CRM:

- *Hardware* refers to the physical equipment in the cockpit including charts, avionic displays, controls, and surfaces.

- *Software* refers to regulations, SOPs, and all the rules that govern aviation activities.

- *Liveware* refers to all of the human beings within the aviation system. This includes pilots, dispatchers, ATC, flight attendants, maintenance staff, ground crew, and even passengers.

Prior to the introduction of CRM, pilot training had advanced very little since World War II. Pilots were trained on the *technical* skills of how to operate an aircraft safely and within specifications. Very little attention was given to *nontechnical* skills such as leadership, teamwork and communication. CRM represented a significant shift in the mindset of the industry, as human factors was legitimized as being important pilot training and much more than just psychobabble.

Since the inception of CRM, improvements have been made to its original focus. For example, CRM training has been expanded to include other airline employees, such as flight attendants, maintenance, and dispatchers. In addition, a spin-off of CRM training has been developed and called *Line Oriented Flight Training* (LOFT). LOFT takes CRM to the next level by creating real-time flight simulator scenarios which are meant to replicate situations that may occur during line operations. These situations are often based on real accident reports and are designed to allow pilots to practice applying CRM skills in real-world situations without the risk associated with practice in an actual aircraft.

CRM is increasingly accepted by all aspects of the industry as essential pilot training. Presently, all major international airlines actively run CRM courses. CRM training is also a component of all military pilot training programs. In a meta-analysis of fifty-eight CRM studies, it was concluded that "CRM training programs seem to produce positive participant reactions, learning, and application of learned behaviour via simulators, on line, or on the job" (Salas, Burke, Bowers, and Wilson 2001 671). CRM training is necessary in aviation because human error is the result of natural human limitations when combined with complicated automated systems. CRM is one of the strategies an organization can use to control error (Helmreich et al. 1999).

CASE

A TRAGIC EXAMPLE OF CRM *FAILURE*

On 29 December 1972 an Eastern Airlines Lockheed L-1011 crashed into the Florida Everglades. Of the 163 passengers and thirteen crew members aboard the aircraft, ninety-four passengers and five crew members were killed in the accident. How could this tragedy have occurred? The first thing that may come to mind is a mechanical failure or perhaps severe weather conditions. In reality this accident was caused by poor communications.

The flight diverted its approach to land because the pilots noticed that the landing gear light did not illuminate to indicate that the nose wheel was locked in the down position. The aircraft climbed to 2000 feet and the pilots began troubleshooting the problem in an effort to determine whether or not the nose wheel was indeed down.

The crew became so distracted by this minor problem that they failed to notice that the autopilot had been disengaged and they began a slow descent towards the ground.

The transcript below is taken from the Cockpit Voice Recorder (CVR) of Eastern Air Lines flight 401.

2342:05, the first officer said, "We did something to the altitude."

The captain's reply was, "What?"

2342:07, the first officer asked, "We're still at two thousand, right." and the captain immediately exclaimed, "Hey, what's happening here?"

2342:10, the first of six radio altimeter warning "beep" sounds began; they ceased immediately before the sound of the initial ground impact.

2342:12, while the aircraft was in a left bank of 28', it crashed into the Everglades at a point 0.7 statute miles west-northwest of Miami International Airport [MIA]. The aircraft was destroyed by the impact.

NTSB 1973 17.

A Tragic Example of *CRM* Success

On 19 July 19 1989 Captain Al Haynes was piloting United flight 232, a DC-10, from Denver to Chicago. During this flight he encountered a situation that was believed to be impossible (or the odds of a billion to 1). The #2 engine experienced an uncontained failure, meaning that, as the engine failed, nearly seventy pieces of shrapnel were thrown through the engine cowl. Some of this shrapnel penetrated the horizontal stabilizer, severing all three hydraulic lines.

The designers of the DC-10 integrated three redundant, completely independent, hydraulic systems and expected that a failure of all three was impossible. The DC-10, like many modern airliners, does not have a mechanical linkage between the control yoke in the cockpit and the flight control surfaces on the wings and tail. The cables from the yoke go to servos, and the hydraulics move the control surfaces.

When United 232 lost all three hydraulic systems, they also lost all aileron, rudder, elevator, spoiler, nose wheel, tail wheel, and brake control. The only means they had to control the aircraft was to manipulate the throttles on the #1 and #3 engine. Through differential thrust, they were able to skid the aircraft into a turn. With a little luck and a lot of skill, Captain Haynes and his crew were able to make an emergency landing on the runway at Sioux City, Iowa. One hundred ten of the 285 passengers and one of the eleven crew members were killed in the crash.

Although this is a tragic story, it is considered an excellent example of Crew Resource Management. In fact, during simulator re-enactments of the event, pilots were unable to exercise any control over the aircraft. The flight crew effectively used all of the available resources to maximize the survivability of the situation. Examples of this include:

- They engaged the help of an off-duty DC-10 flight instructor who was in the cabin riding as a passenger. It was he who determined that some directional control could be accomplished through differential thrust of the engines.

- The captain and first officer worked together effectively to diagnose the problem; communicate with ATC, dispatch, and passengers; control the aircraft; and prepare for landing.

- Air Traffic Control was continually in contact with the aircraft. They were able to point out alternate areas to land, including a four-lane highway, should the aircraft not make the airport. They cleared the traffic at the airport and alerted emergency vehicles to be ready at the far end of the runway.

NTSB Aircraft Accident Report: United Airlines Flight 232, 1990.

Single-Pilot Resource Management

CRM has increased in recognition and acceptance in the aviation industry since its creation in 1979. However, until recent years there was no version of CRM training available for those operating in a single-pilot configuration. The majority of this type of flying occurs in general aviation (GA). In recent years, GA aircraft have experienced a rapid increase in technological sophistication with the introduction of glass cockpit avionic displays. In addition, very light jets (VLJs) are set to revolutionize GA and the aviation industry as a whole. VLJs provide the speed and efficiency of jet-powered flight to moderately experienced pilots at a reasonable cost of ownership. Also, VLJs are certified for single pilot operations (Strait 2006). An example of a VLJ is the Canadian produced D-JET from Diamond Aircraft.

Courtesy Diamond Aircraft.

Figure 7.7

These advancements will result in GA pilots flying faster and more technologically advanced aircraft than previously available. In addition to this, GA pilots are at additional risk, due to the relative inexperience of pilots in this sector compared to military and commercial aviation. Pilots with limited experience are particularly susceptible to accidents and incidents as their confidence level may exceed their level of skill. In particular, two periods are especially dangerous, (a) at approximately 100 hours total time when the pilot has accumulated about fifty hours past their private pilot license/certificate, and (b) between fifty and 100 hours after completing their instrument rating (Trollip and Jensen 1991). Typically, both of these high risk periods will take place while a pilot is operating in the GA sector.

In response to these factors, the industry recognized the need for CRM concepts to be taught in single-pilot operators. This resulted in the development of a single pilot resource management (SRM) training overview (FITS n.d.; NBAA Safety Committee 2005). However, SRM training is still its infancy compared to CRM, and this term is not yet well known. Based on the descriptions formulated by the FAA and NBAA, SRM consists of nontechnical training with a goal of reducing pilot error. The major distinction between SRM and CRM is that air crew coordination and communication, major components of CRM training, are not included in SRM (FITS n.d.; NBAA Safety Committee 2005). However, other CRM components are adapted for single pilot operations and included in SRM. These include:

- Communication: Instead of between the captain and first officer, SRM communication focuses on the pilot's interaction with air traffic controllers, dispatch, and other aircraft.

- Situation Awareness: This is a crucially important safety concept, as one investigation revealed that human error was the primary causal factor in 97 percent of GA accidents and that a significant number of these were the result of situation awareness (NTSB 1989; Proctor, Panko, and Donovan 2004). The definition of situation awarness is the ability to perceive elements in the pilot's environment, understand what those elements are, and predict how they will impact the pilot in the near future (Endsley 1995).

- Aeronautical decision making: Refers to the pilot's ability to make effective in-flight decisions by considering all available information and outcomes.

- Workload management: The workload of pilots involves a continual stream of mental and physical tasks. Workload management training typically focuses on teaching pilots strategies they can use to better prioritize multiple tasks simultaneously. The A-N-C-S hierarchy is the task prioritization model of aviate, navigate, communicate, and systems management. It helps pilots remember that the most important task is to fly the plane! After this is stabilized, attention can be focused on navigating, communicating, and managing onboard systems.

- Automation management: Modern flight decks are becoming increasingly automated. A pilot's ability to manage automation leads directly to the safety of the flight. There are many issues that a pilot can have with this task, such as automation complacency. When a pilot becomes complacent with automation, they hand over all responsibility for the flight. This may work in the short-term, but if the automation ever fails (which it tends to do in the most difficult and complicated circumstance), the pilot is so far out of the loop that it is very difficult to respond effectively.

All of these factors are important to pilot safety.

Western News.

Figure 7.8 *A pilot completes online single-pilot resource management (OSRM) training.*

Research into the feasibility of computer-based SRM has demonstrated that a ninety-minute training program can significantly improve a pilot's situation awareness (Kearns 2007). This is a very important finding, as it allows SRM training to be delivered to GA pilots who are distributed across the country. Since its introduction in 2008, online single-pilot resource management (OSRM) has been completed by hundreds of pilots across Canada.

Conclusion

Overall, human factors are crucially important in aviation. With an understanding of human factors, an aviation professional can develop an understanding of not only what is going wrong, but **why** it is going wrong. This understanding allows for the use of one of the five approaches to address human factors errors (equipment, environment, task, selection, and/or training) to eliminate the problem.

Making mistakes is a part of being a human being. It is not feasible to expect error-free performance in any system where a human is involved in the operation, manufacture, or maintenance. What **can** be done is to develop an understanding of how humans interact with the system, keeping in mind natural human limitations. This can allow for the identification of risks before an accident occurs.

Name _____ **Date** _____

PRACTICE ACTIVITY

Review the following photographs. Without fully knowing the background of the situation, describe in the space allotted how you think human error might have contributed to the situation. Then, ask yourself which of the five human factors approaches could be used to solve the problem (equipment, task, environment, selection, and/or training). There is no right or wrong answer. This activity is meant to give you some practice in applying human factors approaches. In the space provided, describe how the approach you chose could prevent similar situations from occurring again.

Describe how human error could have contributed to this situation.

_____

_____

_____

Which human factors approach would you implement to prevent future occurrences?

_____

Describe how you would implement this approach and how it may prevent future occurrences of the situation.

_____

_____

_____

Describe how human error could have contributed to this situation.

_____

_____

_____

Which human factors approach would you implement to prevent future occurrences?

_____

Describe how you would implement this approach and how it may prevent future occurrences of the situation.

_____

_____

_____

Describe how human error could have contributed to this situation.

_____

_____

_____

Which human factors approach would you implement to prevent future occurrences?

_____

Describe how you would implement this approach and how it may prevent future occurrences of the situation.

_____

_____

_____

Describe how human error could have contributed to this situation.

_____

_____

Which human factors approach would you implement to prevent future occurrences?

_____

Describe how you would implement this approach and how it may prevent future occurrences of the situation.

_____

_____

_____

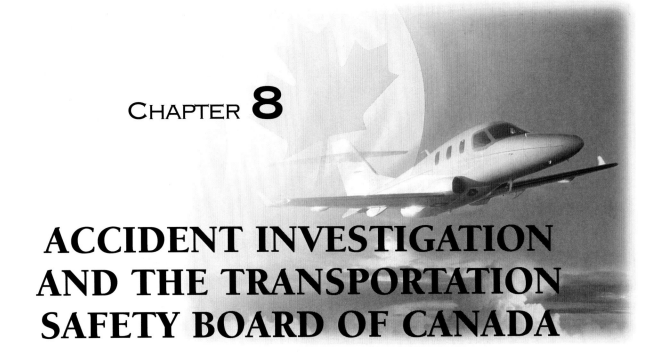

CHAPTER **8**

ACCIDENT INVESTIGATION AND THE TRANSPORTATION SAFETY BOARD OF CANADA

OBJECTIVES

After you have read this chapter, you will understand:

- The history, purpose, and organization of the Transportation Safety Board of Canada (TSB)

- Career options within the TSB

- The accident investigation process, including details of the field, post-field, and report production phases

- The definition of an aviation "accident" versus an aviation "incident"

- TSB's SECURITAS voluntary, confidential, non-punitive incident reporting system

History and Organization of the TSB

On 29 March 1990, the *Canadian Transportation Accident Investigation and Safety Board Act* (CTAISB Act) came into force. This Act of Parliament was responsible for the creation of the Transportation Safety Board (TSB) of Canada. The Transportation Safety Board of Canada (TSB) is an independent agency responsible for investigating marine, pipeline, railway, and aviation transportation occurrences. Its sole purpose is to advance transportation safety in Canada. The TSB reports directly to Parliament through the President of the Queen's Privy Council for Canada. It is entirely separate from Transport Canada (the transportation regulatory authority in Canada). The TSB is headed by a Chairperson and up to four Board members. Its Head Office is located in Gatineau, Quebec, with an extensive engineering laboratory in Ottawa, Ontario. A little more than one-third of the 230 employees within the TSB are located in field and regional offices across Canada. This distribution of personnel allows a rapid response to emergency situations that occur anywhere across the country.

THE MANDATE OF THE TSB IS TO ADVANCE SAFETY IN THE MARINE, PIPELINE, RAIL, AND AIR MODES OF TRANSPORTATION BY:

(A) CONDUCTING INDEPENDENT INVESTIGATIONS, INCLUDING PUBLIC INQUIRIES WHEN NECESSARY, INTO SELECTED TRANSPORTATION OCCURRENCES IN ORDER TO MAKE FINDINGS AS TO THEIR CAUSES AND CONTRIBUTING FACTORS

(B) IDENTIFYING SAFETY DEFICIENCIES, AS EVIDENCED BY TRANSPORTATION OCCURRENCES

(C) MAKING RECOMMENDATIONS DESIGNED TO ELIMINATE OR REDUCE ANY SUCH SAFETY DEFICIENCIES

(D) REPORTING PUBLICLY ON OUR INVESTIGATIONS AND ON THE FINDINGS IN RELATION THERETO.

Part of the investigative process involves monitoring developments in transportation safety and identifying risks that government and companies should address to maximize safety.

To instill confidence in the public regarding the transportation accident investigation process, it was considered essential that the TSB be independent and, therefore, free from potential conflicts of interest in the investigation process. Therefore, Parliament constituted the TSB as an independent agency which is entirely separate from other government departments and agencies. The TSB reports to Parliament through the President of the Queen's Privy Council for Canada. The TSB may work in cooperation with Transport Canada (TC), the Royal Canadian Mounted Police (RCMP), or the Department of National Defense (DND) in the investigation of an accident. The TSB does not investigate aviation accidents that occur in the military.

TSB's independence allows for complete objectivity in the development of findings as to what caused an accident and safety recommendations. When an accident is investigated by the TSB, Transport Canada is not permitted to investigate for the purpose of identifying causal and contributing factors to the accident. Transport Canada may investigate for the separate purpose of identifying regulatory infractions.

The purpose of the TSB's investigative work is to determine what caused and contributed to the accident, **not** to assign blame or liability. Due to the nature of this work, it is common for legal action to be taken against those who may be identified as contributing to the accident. However, the statute is clear in stating that no finding of the Board shall be construed as assigning fault or determining civil or criminal liability. Consequently, the TSB will **not** refrain from reporting on the cause and contributing factor to an accident because fault or liability may be inferred.

Organization of TSB

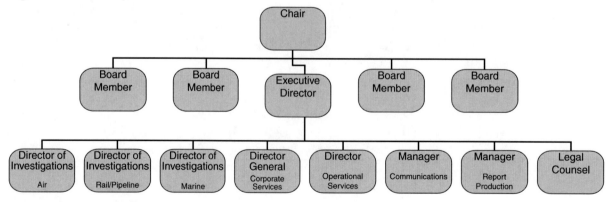

Transportation Safety Board, 2009.

Figure 8.1 *Organizational Structure of the Transportation Safety Board.*

There is a Chairperson of the Board and up to four Board members who have some background in one of the four transportation categories. The Executive Director reports directly to the Chairperson as the Board has no management oversight responsibilities. Reporting to the Executive Director are branches responsible for communications, operational services, and corporate services, as well as three modally oriented investigative branches (air, marine, and rail/pipeline) and a legal counsel. Within the air investigation sector, there are operational, technical, air traffic control, and engineering investigators.

PROFESSIONAL PROFILE

Job Title:

Transportation Safety Board Accident Investigator (operational and technical)

Educational and Work Experience:

Operational investigators have over 10,000 hours of pilot experience, typically an airline transport pilot's license or a commercial helicopter pilot's license with an instrument rating. The TSB also looks for candidates to be a captain and to have some management experience.

Technical investigators must have an aircraft maintenance engineer (AME) license. They need many years experience in the industry with some supervisory or management experience.

Due to the expertise required for this occupation, this is not a job that people typically begin before the age of thirty. Once hired as an investigator, one must complete a basic aircraft accident investigation course. Throughout one's career, supplemental training is typically required. Such training courses may advance aircraft accident investigation, gas turbine investigation, helicopter investigation, human factors, accident photography, witness interviewing, and media relations.

Salary:

The salary range for operational investigators is $89,500–103,000/year, and the technical investigators salary is $77,000–87,200/year.

Professional Characteristics:

People in this line of work typically have the high moral standing of wanting to improve safety. This occupation has excellent stability and regular nine-to-five hours (compared to work as a pilot or AME). In the investigation of fatal accidents, investigators will be required to witness, evaluate, and interact with human remains. This can occasionally be challenging and stressful. Overall, however, this is a job that investigators truly enjoy, as they get the opportunity to directly impact safety.

TSB Investigations

Every year, roughly 4000 transportation occurrences are reported to the TSB. Formal Investigations are conducted in approximately 2–3 percent of reported occurrences each year. For the remainder, the TSB collects data on the pertinent circumstances for future research or use as required to advance safety. When the TSB is notified of an occurrence it will evaluate whether or not an investigation is required. This assessment often includes deploying an investigation team to the site of the occurrence. One of the determining factors in whether or not to launch a full investigation is based on a probability assessment of whether an investigation will reduce future risks to people, property, or the environment.

When Canadian interests are involved, the TSB occasionally aids in foreign investigations. For example, if an aircraft that is registered, licensed, or manufactured in Canada has an accident internationally, the TSB will participate in the investigation, along with the investigating authorities of the country of occurrence.

TSB investigators share stand-by duty. Typically, for one week at a time, an investigator will wear a pager and must be available twenty-four hours a day to respond to an emergency call. The investigator on stand-by duty will often be assigned as the investigator-in-charge (IIC), should an accident occur during his tour of duty. Usually, investigators are deployed as a team of two with one technical investigator and one operational investigator.

TSB's Investigation Process

There are three main phases of a TSB investigation: field, post-field, and report production.

Field Phase

Once the decision has been made to investigate an occurrence, an investigator-in-charge (IIC) is appointed and an investigation team is assembled. The makeup of the team is based on the circumstances of the occurrence and investigation needs, and may consist of operations, equipment, maintenance, engineering, scientific, and human performance specialists. The number of investigators sent to an occurrence site varies from one or two, for a relatively straightforward investigation, to ten or perhaps many more for a major investigation.

The field phase can last from one day to several months. During the field phase, team members generally:

- Secure and examine the occurrence site
- Examine the equipment, vehicle, or wreckage
- Interview witnesses and company and government personnel
- Collect pertinent information
- Select and remove specific wreckage items for further examination
- Review documentation
- Identify potential unsafe acts and unsafe conditions

Transportation Safety Board of Canada

Figure 8.2 *TSB investigators at the accident scene following the 14 October 2004 crash of an MK Airlines Boeing 747 in Halifax, NS.*

Post-Field Phase

A great deal of work remains when the investigation team returns from the accident site. This phase may take several months. During this phase, the TSB investigators will sort through all of the data gathered to identify safety significant events. Activities in this phase include:

- Examining all records associated with the aircraft, pilot, and aviation company involved

- Examining wreckage, systems, and components in the laboratory

- Analyzing cockpit voice recorder and flight data recorder reports

- Using computer-based reconstruction to simulate the accident and develop a timeline of events

- Considering pilot toxicology and autopsy reports

- Conducting additional interviews

When safety deficiencies are suspected or confirmed, the TSB will advise the appropriate person or authority as quickly as possible, without waiting until publication of the final report, so that the problem can be corrected. The Board may make formal recommendations to draw immediate attention to particular safety deficiencies, especially if it has been determined that the risk of recurrence of a similar accident is high and its consequences severe. The TSB practice is not to wait for a final report or until an investigation is complete to make important safety information public.

At the end of the post-field phase, the IIC produces an initial draft investigation report.

Report Production Phase

This phase is the most time-consuming of the three accident investigation phases. During the report production phase, the Board reviews the initial draft investigation report, which may be approved, amended, or returned for further staff work. Once the initial draft report is approved by the Board, it is sent as a confidential draft report to persons and corporations whose interests may be affected by the report and who are most qualified to comment on its accuracy. They then have the opportunity to dispute, correct, or contradict information that they believe is incorrect or unfairly prejudicial to their interests.

This process is intended to ensure procedural fairness and the accuracy of the Board's final report. The Board considers all representations (comments) and will amend the report if required. Once the Board approves the final report, it is translated into both official languages and prepared for release to the public.

The Board aims to release final accident reports as quickly as possible. However, the complexity of the accident may affect the time required to finalize a report. The TSB takes the time necessary to do a thorough investigation and to produce a report that advances safety and meets the expectations of the Canadian public and the transportation industry.

Cooperation

The TSB does not work in isolation. Many people and groups work with the TSB in the investigation of accidents. For example, in a commercial aviation accident, you could expect all levels of government, the aircraft manufacturer, engine manufacturer, airline, witnesses, survivors, next of kin, medical examiners, police, fire departments, and search-and-rescue teams to be involved. The TSB would not be capable of investigating an accident without the full cooperation of a broad network of groups and individuals.

Police officers are usually the first to arrive at the scene of an accident. It is their responsibility to secure the site, preserve evidence, document the situation, assist the coroner, and look for signs of criminal activity.

An unfortunate aspect of this line of work is that TSB investigators must deal with fatalities on a regular basis. However, human remains can sometimes provide important clues regarding what caused the accident. When TSB investigators arrive at a fatal accident site, the bodies of those onboard the aircraft are typically still inside the fuselage. In this situation, a coroner or medical examiner is present at the accident site. The

TSB investigator and coroner jointly manage the accident site until the deceased have been removed. The coroner's responsibility is to remove the deceased and determine the cause of death. Once this has occurred, the authority over the accident is normally relegated to the TSB, with the concurrence of the police forces present.

The TSB is not responsible for notifying the next of kin of an accident. This responsibility lies with the police, coroner, or transportation company (such as the airline). However, the TSB plays a major role in communicating with next of kin to keep them abreast of how the investigation is progressing. Additionally, the TSB may request interviews with next of kin to assist in the investigation process. It is common for interviewees to be joined by a family member, lawyer, friend, or union representative during the interview. No names are released in the final TSB report.

Observers

It is fairly common for the Minister of Transport or any other minister to have a direct interest in the subject matter of an investigation. Therefore, a minister may, in accordance with subsection 23(2) of the CTAISB Act, designate a person to attend as an observer. The Minister's Observer will obtain factual information from an ongoing investigation, advise the Department of any significant regulatory factors, identify deficiencies that require immediate corrective actions, and coordinate the required support to an occurrence investigation. The Minister's Observer will not participate in any regulatory investigation or enforcement action taken by the Department in relation to any occurrence to which the observer is assigned.

The TSB may grant observer status to persons with a direct interest (PDI) in the investigation at its own discretion, if it is felt they may contribute positively to the investigation. PDIs may be individuals from airlines, aircraft or engine manufacturers, foreign regulating agencies, or any individuals the TSB believes may contribute to the success of the investigation. PDIs are supervised by a TSB investigator.

Investigation Reports

Every investigation is given an occurrence file number to allow for quick and easy dissemination of information. The TSB may release factual information about the occurrence throughout the investigation. Information that may impact safety is shared immediately with any individuals or groups that may implement changes to improve safety. This information may be in the form of recommendations, safety advisories, safety information letters, or in any format that will allow the safety message to be passed, so that corrections may be applied without delay.

However, the nature of the investigatory process is that some of the information is sensitive in nature. This information, such as cockpit voice recordings and witness statements, is protected by law. In Canada, these items are protected under the Privacy Act and the CTAISB Act; therefore, the recordings and transcripts will not be publically released.

Definition of Accident

Before one can begin studying safety statistics (as presented in Chapter 9), it is crucial to have an understanding of what type of occurrence constitutes a reportable accident compared to a reportable incident. The term *reportable* means that reporting the occurrence to the Transportation Safety Board is mandatory and not at the discretion of the people involved.

ACCORDING TO THE TRANSPORTATION SAFETY BOARD REGULATIONS:

A "REPORTABLE AVIATION ACCIDENT" MEANS AN ACCIDENT RESULTING DIRECTLY FROM THE OPERATION OF AN AIRCRAFT, WHERE

(A) A PERSON SUSTAINS A SERIOUS INJURY OR IS KILLED AS A RESULT OF

    A. BEING ON BOARD THE AIRCRAFT,

    B. COMING INTO CONTACT WITH ANY PART OF THE AIRCRAFT OR ITS CONTENTS, OR

    C. BEING DIRECTLY EXPOSED TO THE JET BLAST OR ROTOR DOWNWASH OF THE AIRCRAFT

(B) THE AIRCRAFT SUSTAINS DAMAGE OR FAILURE THAT ADVERSELY AFFECTS THE STRUCTURAL STRENGTH, PERFORMANCE OR FLIGHT CHARACTERISTICS OF THE AIRCRAFT AND THAT REQUIRES MAJOR REPAIR OR REPLACEMENT OF ANY AFFECTED COMPONENT PART, OR

(C) THE AIRCRAFT IS MISSING OR INACCESSIBLE.

TO AID IN UNDERSTANDING THIS REGULATION, IT IS HELPFUL TO BREAK IT DOWN INTO COMPONENTS. FIRST, AN OCCURRENCE IS CONSIDERED AN "ACCIDENT" IF A PERSON SUSTAINS A SERIOUS INJURY OR IS KILLED. A SERIOUS INJURY IS DEFINED IN THE TSB REGULATIONS AND THE CANADIAN AVIATION REGULATIONS (CARs) AS AN INJURY THAT IS LIKELY TO REQUIRE ADMISSION TO A HOSPITAL. THIS INJURY OR DEATH MUST OCCUR ON BOARD THE AIRCRAFT, AFTER COMING INTO CONTACT WITH THE AIRCRAFT, OR AFTER BEING EXPOSED TO JET BLAST OR ROTOR DOWNWASH. AS AN EXAMPLE, A PASSENGER OR CREW MEMBER WHO IS INJURED OR KILLED AS A RESULT OF COMING INTO CONTACT WITH THE PROPELLER OF AN AIRCRAFT OR THE ROTOR OF A HELICOPTER THAT IS INTENDED TO GO FLYING IS A REPORTABLE ACCIDENT.

THE SECOND ASPECT OF THIS REGULATION DEFINES A REPORTABLE ACCIDENT AS ANY TIME AN AIRCRAFT SUSTAINS DAMAGE OR FAILURE THAT AFFECTS ITS ABILITY TO FLY AND REQUIRES REPAIR OR REPLACEMENT. THIS CATEGORY, WHICH INCLUDES MECHANICAL FAILURES AND MAJOR DAMAGE ASSOCIATED WITH CRASHES, IS MORE TYPICAL OF WHAT IS EXPECTED WHEN CONSIDERING AVIATION ACCIDENTS.

LAST, IF THE AIRCRAFT CANNOT BE FOUND OR IS INACCESSIBLE, IT IS CONSIDERED A REPORTABLE AVIATION ACCIDENT. THE TSB DOES NOT WAIT FOR AN AIRCRAFT TO BE LOCATED BEFORE BEGINNING AN INVESTIGATION. THIS IS IMPORTANT AS, EVEN IN MODERN-DAY AVIATION, SOME AIRCRAFT GO MISSING AND ARE NEVER FOUND (OR ARE FOUND YEARS LATER).

THE LOST ADVENTURER

Steve Fossett was an American millionaire, aviator, businessman, and adventurer. He was the first pilot to fly solo nonstop around the world in a lighter-than-air balloon. He held approximately 100 world records at the time of his death, mostly in sailing and aviation. On 3 September 2007, he took a rented Citabria Super Decathlon aircraft for a flight from a private airstrip in Nevada. He was reported missing on 4 September and a massive search-and-rescue initiative was initiated. Ironically, the search-and-rescue teams uncovered wreckage of at least eight other aircraft in western Nevada but were unable to locate Mr. Fossett's aircraft. On 15 February 2008, a Cook County judge determined there was sufficient evidence to declare Mr. Fossett dead. On 1 October 2008, a hiker found three identification cards in the desert, along with some money. On 2 October 2008, search teams in California sighted what appeared to be aircraft wreckage. Investigators stated that it appeared Mr. Fossett's aircraft had struck the mountainside head-on, disintegrating most of the fuselage. Later, bones were found, and a forensics lab determined that DNA matched that of Steve Fossett (BBC 2008).

Definition of Incident

An aviation incident can be considered a near-accident. An unsafe situation occurred, but it was less serious than an accident (no one was seriously injured or killed, the aircraft was not substantially damaged, and the aircraft was not missing). Although some may consider incidents less important to investigate than accidents, these occurrences must not be overlooked. When incidents are investigated, it may be possible to identify risk factors and eliminate them before an accident occurs. From a safety perspective, incidents involve the same safety data as accidents with the exception of the one terminal event or situation that would have escalated the incident to a major accident. However, there are several advantages to investigating incidents including:

- People are alive and able to relay information about the occurrence.

- It is less likely that legal action will be brought forward, so more people are willing to provide interviews.

- There is less media scrutiny.

A "REPORTABLE AVIATION INCIDENT" MEANS AN INCIDENT RESULTING DIRECTLY FROM THE OPERATION OF AN AIRPLANE HAVING A MAXIMUM CERTIFICATED TAKE-OFF WEIGHT GREATER THAN 5700 KG, OR FROM THE OPERATION OF A ROTORCRAFT HAVING A MAXIMUM CERTIFICATED TAKE-OFF WEIGHT GREATER THAN 2250 KG, WHERE

(A) AN ENGINE FAILS OR IS SHUT DOWN AS A PRECAUTIONARY MEASURE,

(B) A TRANSMISSION GEARBOX MALFUNCTION OCCURS,

(C) SMOKE OR FIRE OCCURS,

(D) DIFFICULTIES IN CONTROLLING THE AIRCRAFT ARE ENCOUNTERED OWING TO ANY AIRCRAFT SYSTEM MALFUNCTION, WEATHER PHENOMENA, WAKE TURBULENCE, UNCONTROLLED VIBRATIONS OR OPERATIONS OUTSIDE THE FLIGHT ENVELOPE,

(E) THE AIRCRAFT FAILS TO REMAIN WITHIN THE INTENDED LANDING OR TAKE-OFF AREA, LANDS WITH ALL OR PART OF THE LANDING GEAR RETRACTED OR DRAGS A WING TIP, AN ENGINE POD OR ANY OTHER PART OF THE AIRCRAFT,

(F) ANY CREW MEMBER WHOSE DUTIES ARE DIRECTLY RELATED TO THE SAFE OPERATION OF THE AIRCRAFT IS UNABLE TO PERFORM THE CREW MEMBER'S DUTIES AS A RESULT OF A PHYSICAL INCAPACITATION THAT POSES A THREAT TO THE SAFETY OF ANY PERSON, PROPERTY OR THE ENVIRONMENT,

(G) DEPRESSURIZATION OCCURS THAT NECESSITATES AN EMERGENCY DESCENT,

(H) A FUEL SHORTAGE OCCURS THAT NECESSITATES A DIVERSION OR REQUIRES APPROACH AND LANDING PRIORITY AT THE DESTINATION OF THE AIRCRAFT,

(I) THE AIRCRAFT IS REFUELLED WITH THE INCORRECT TYPE OF FUEL OR CONTAMINATED FUEL,

(J) A COLLISION, A RISK OF COLLISION OR A LOSS OF SEPARATION OCCURS,

(K) A CREW MEMBER DECLARES AN EMERGENCY OR INDICATES ANY DEGREE OF EMERGENCY THAT REQUIRES PRIORITY HANDLING BY AN AIR TRAFFIC CONTROL UNIT OR THE STANDING BY OF EMERGENCY RESPONSE SERVICES,

(L) A SLUNG LOAD IS RELEASED UNINTENTIONALLY OR AS A PRECAUTIONARY OR EMERGENCY MEASURE FROM THE AIRCRAFT, OR

(M) ANY DANGEROUS GOODS ARE RELEASED IN OR FROM THE AIRCRAFT.

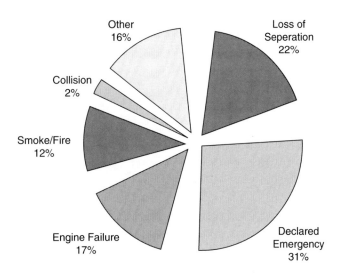

TSB 2007.

Table 8.1 *Types of Reportable Incidents in Canada*

SECURITAS

Beyond reportable accidents and incidents, other types of unsafe situations occur in aviation. Although it is not mandatory to report other types of unsafe situations, SECURITAS was launched in 1985 to collect voluntary reports. SECURITAS is a confidential, voluntary, and non-punitive reporting system that any person can use to submit details of unsafe acts or conditions relating to Canadian aviation. The benefits of submitting a report are that this data may support TSB investigations and analyses on safety matters, including operating procedures, training, human factors, and equipment stability. In addition, through analysis of these reports, widespread safety deficiencies may be identified. This may allow for accidents to be prevented before they occur.

To submit a confidential report to SECURITAS:

Write, fax, e-mail or telephone SECURITAS at

SECURITAS
PO Box 1996
Station B
Gatineau, Quebec
J8Z 3Z2

Tel.: 1-800-567-6865
Fax: (819) 994-8065
E-mail: securitas@bst-tsb.gc.ca

1. Include the following information in your message:

   • Your name, address and phone number;

   • Your profession and experience;

   • Your involvement in the unsafe situation being reported;

   • Where else you have reported this unsafe situation or safety concern;

   • Complete identification of the aircraft, ship or rolling stock; and

   • Owner/operator of the equipment.

2. Describe (as appropriate) the unsafe act or safety concern in terms of

   • how the unsafe act/condition was discovered; and

   • if you are describing an event, what happened, where, when (give the date of the event and local time) and why you think it occurred.

3. Give your suggestions to correct the situation.

Confidentiality is an important aspect of voluntary reporting to Securitas. Without a guarantee of confidentiality, persons may be reluctant to report safety deficiencies out of fear of embarrassment or losing their job. To ensure confidentiality, TSB regulations prohibit the release of any information that could potentially reveal a reporter's identity. All SECURITAS correspondence is sent directly into the SECURITAS office and handled only by authorized SECURITAS analysts. Any information that may potentially be used to identify the reporter is deleted after the report has been reviewed.

Depending on the severity of the hazard reported, the TSB may initiate safety action by issuing a Board recommendation or another type of safety communication aimed at eliminating or reducing the risk posed by the hazard. However, it typically takes several reports to identify and validate a systemic safety deficiency. Several reports of similar situations, coupled with accident/incident data from external sources, are used to identify unsafe conditions that need to be addressed, in view of the risk they pose and their consequence.

Once the information in a SECURITAS report is validated as a safety deficiency, the TSB typically forwards the information with a suggested corrective action to Transport Canada or any stakeholder who is in the best position to fix the problem. No action will ever be taken that might compromise a reporter's identity.

ASRS

In the United States a similar voluntary, confidential, non-punitive reporting system called the Aviation Safety Reporting System (ASRS) is in place. ASRS was implemented as a joint project of the National Aeronautic and Space Administration (NASA) and the Federal Aviation Administration (FAA) in April of 1976. ASRS allows for online reports to be submitted by pilots, air traffic controllers, maintenance technicians, flight attendants, dispatchers, and other aviation employees who have been witness to or involved in an unsafe situation. Like SECURITAS, the purpose of ASRS is to reduce the likelihood of aviation accidents. However, the particular concern of ASRS is the quality of human performance in aviation.

Since 1976, ASRS has received over 723,000 incident reports and spawned sixty major research studies. The general public now has direct access to the ASRS database through the Internet. This service allows pilots in Canada to search for safety incidents in the United States. This can be a valuable tool for Canadian pilots who are wondering how widespread an unsafe situation may be. The ASRS group takes great pride in the fact that they have never revealed the source of a report.

DO NOT REPORT AIRCRAFT ACCIDENTS AND CRIMINAL ACTIVITIES ON THIS FORM.
ACCIDENTS AND CRIMINAL ACTIVITIES ARE NOT INCLUDED IN THE ASRS PROGRAM AND SHOULD NOT BE SUBMITTED TO NASA.
ALL IDENTITIES CONTAINED IN THIS REPORT WILL BE REMOVED TO ASSURE COMPLETE REPORTER ANONYMITY.

IDENTIFICATION STRIP: *Please fill in all blanks to ensure return of strip.*
NO RECORD WILL BE KEPT OF YOUR IDENTITY. This section will be returned to you.

(SPACE BELOW RESERVED FOR ASRS DATE/TIME STAMP)

TELEPHONE NUMBERS where we may reach you for further details of this occurrence:

HOME Area _____ No. _____ Hours _____

WORK Area _____ No. _____ Hours _____

NAME _____

ADDRESS/PO BOX _____

_____

CITY _____ STATE _____ ZIP _____

TYPE OF EVENT/SITUATION _____

DATE OF OCCURRENCE _____

LOCAL TIME (24 hr. clock) _____

PLEASE FILL IN APPROPRIATE SPACES AND CHECK ALL ITEMS WHICH APPLY TO THIS EVENT OR SITUATION.

REPORTER	FLYING TIME	CERTIFICATES/RATINGS	ATC EXPERIENCE
☐ Captain ☐ First Officer 　☐ pilot flying 　☐ pilot not flying ☐ Other Crewmember ☐ _____	total _____ hrs. last 90 days _____ hrs. time in type _____ hrs.	☐ student　☐ private ☐ commercial　☐ ATP ☐ instrument　☐ CFI ☐ multiengine　☐ F/E ☐ _____	☐ FPL　☐ Developmental radar _____ yrs. non-radar _____ yrs. supervisory _____ yrs. military _____ yrs.

AIRSPACE	WEATHER	LIGHT/VISIBILITY	ATC/ADVISORY SERV.
☐ Class A (PCA) ☐ Class B (TCA) ☐ Class C (ARSA) ☐ Class D (Control Zone/ATA) ☐ Class E (General Controlled) ☐ Class G (Uncontrolled) ☐ Special Use Airspace ☐ airway/route _____ ☐ unknown/other _____	☐ VMC　☐ ice ☐ IMC　☐ snow ☐ mixed　☐ turbulence ☐ marginal ☐ tstorm ☐ rain　☐ windshear ☐ fog　☐	☐ daylight　☐ night ☐ dawn　☐ dusk ceiling _____ feet visibility _____ miles RVR _____ feet	☐ local　☐ center ☐ ground　☐ FSS ☐ apch　☐ UNICOM ☐ dep　☐ CTAF Name of ATC Facility: _____

	AIRCRAFT 1	AIRCRAFT 2
Type of Aircraft (Make/Model)	(Your Aircraft) _____　☐ EFIS　☐ FMS/FMC	(Other Aircraft) _____　☐ EFIS　☐ FMS/FMC
Operator	☐ air carrier　☐ military　☐ corporate ☐ commuter　☐ private　☐ other _____	☐ air carrier　☐ military　☐ corporate ☐ commuter　☐ private　☐ other _____
Mission	☐ passenger　☐ training　☐ business ☐ cargo　☐ pleasure　☐ unk/other_____	☐ passenger　☐ training　☐ business ☐ cargo　☐ pleasure　☐ unk/other_____
Flight plan	☐ VFR　☐ SVFR　☐ none ☐ IFR　☐ DVFR　☐ unknown	☐ VFR　☐ SVFR　☐ none ☐ IFR　☐ DVFR　☐ unknown
Flight phases at time of occurrence	☐ taxi　☐ cruise　☐ landing ☐ takeoff　☐ descent　☐ missed apch/GAR ☐ climb　☐ approach　☐ other _____	☐ taxi　☐ cruise　☐ landing ☐ takeoff　☐ descent　☐ missed apch/GAR ☐ climb　☐ approach　☐ other _____
Control status	☐ visual apch　☐ on vector　☐ on SID/STAR ☐ controlled　☐ none　☐ unknown ☐ no radio　☐ radar advisories	☐ visual apch　☐ on vector　☐ on SID/STAR ☐ controlled　☐ none　☐ unknown ☐ no radio　☐ radar advisories

If more than two aircraft were involved, please describe the additional aircraft in the "Describe Event/Situation" section.

LOCATION	CONFLICTS
Altitude _____ ☐ MSL ☐ AGL **Distance and radial from airport, NAVAID, or other fix** _____ _____ **Nearest City/State** _____	Estimated miss distance in feet: horiz _____ vert _____ Was evasive action taken?　☐ Yes　☐ No Was TCAS a factor?　☐ TA　☐ RA　☐ No Did GPWS activate?　☐ Yes　☐ No

NASA ARC 277B (January 1994)　　**GENERAL FORM**　　Page 1 of 2

http://asrs.arc.nasa.gov/.

Figure 8.2 *The ASRS form has been designed to specifically gather the maximum amount of human factors data without overwhelming the reporter. All identification information is deleted before the report is entered into the computer. A second page (not pictured) is included in the ASRS report which allows the reporter to write in a description of the event.*

172 Accident Investigation and the Transportation Safety Board of Canada

Swissair Flight 111

On 2 September 1998, Swissair 111, a McDonnell Douglas MD-11, crashed into the ocean off of Peggy's Cove, Nova Scotia. All 215 passengers and fourteen crew members were killed in the accident. The TSB led the investigation to determine what caused the accident.

Overall, this was perhaps the biggest and most complex investigation in aviation history. This accident cost $57 million and took five years to complete.

Mr. Peter Rowntree, a senior regional investigator with the TSB, offered the following insight about his involvement in the investigation.

"I still have nightmares about what I saw during the Swissair investigation."

Several investigators relocated to Nova Scotia during the process of the investigation, separated from friends and family and living in a hotel in the area for years at a time.

When the aircraft hit the water, the destruction was so extensive that it broke into millions of pieces, and each one needed to be first retrieved from the ocean and then looked at by hand. Along with the retrieved aircraft wreckage, bodies of passengers needed to be salvaged to confirm death. This retrieval was particularly difficult, as most could be identified only by dental records, fingerprints, and DNA.

Near the end of the recovery phase, the TSB contracted the ship *Queen of the Netherlands,* the largest suction-dredging ship in the world. They dredged the ocean floor two metres deep across an area the size of two football fields. Everything was pumped into a holding cell, the water and sand were removed, and machinery was used to separate all matter from the aircraft wreckage.

TSB 2007.

Figure 8.3 *Barge used for Swissair Flight 111 wreckage recovery.*

Once wreckage was retrieved, investigators examined every single piece (refer to the photo). Manufacturers and others, such as the RCMP, assisted in identifying the pieces of wreckage. A temporary hangar was established to house the millions of pieces of wreckage, and family tours were constantly given. Other investigators worked on the CVR and FDR analysis every day. This was particularly challenging because both recorders lost power six minutes before the aircraft's impact.

TSB 2007.

Figure 8.4 *TSB investigators examining wreckage of Swissair Flight 111, which crashed off of Peggy's Cove, NS, on 2 September 1998.*

Therefore, the last six minutes of the flight had to be reconstructed through analysis of the wreckage. To accomplish this monumental task, investigators built a metal skeleton of the front ten metres of the aircraft. When pieces were identified as coming from this area of the aircraft, they were reconstructed on the skeleton (refer to the photo). In this manner, investigators were able to identify an area in the ceiling area above the cockpit where an in-flight entertainment system wire arced and set fire to insulation coverings that were supposed to be fire resistant.

Figure 8.5 *Overhead view of full-scale flight deck reconstruction, Swissair Flight 111.*

Overall, the TSB identified eleven causal and contributing factors to the Swissair 111 accident, and another twenty-four findings were made as to risk. As a result of this investigation, the Board made twenty-three recommendations to the aviation industry with the purpose of avoiding accidents such as this in the future. Although most of these recommendations have been adopted within the aviation industry, some have not. It can be somewhat frustrating when some recommendations are not implemented. However, the perspective of the TSB is to promote safety. Sometimes the recommendations can't be adopted immediately because of the cost of the changes or a lack of technology/replacement material available.

The TSB "doesn't have the power to force people to make changes. We can identify safety deficiencies but we can't prescribe the fix."

Conclusion

The Transportation Safety Board (TSB) has an international reputation as one of the best aviation accident investigation bodies in the world. This reputation has been built on the outstanding work they have done on complicated and high-profile investigations such as Swissair 111. In these investigations, the world is watching and many parties have a stake in the findings. Through their impartiality, professionalism, humanity, and efficiency, the TSB members are an important contributor to the fabric of aviation safety in Canada.

PRACTICE ACTIVITY

Consider the following situation:

You are a pilot who has been flying at Joe & Bob's Flight School for the last three years. Twelve months ago, Joe sold his share in the company to Bob and moved to the Bahamas. Bob is now the general manager and chief flight instructor at the school. Since that time, you have noticed that maintenance and customer service standards have declined steadily. Although all of the records seem to be in order, you suspect that management is cutting corners. In addition, anyone in the company who has voiced concerns to Bob (including your last flight instructor) has been fired from the company and replaced with inexperienced staff. No one at the school is willing to speak up about these low standards because they are afraid of losing their jobs. On 19 March, you rented aircraft C-XXXX for an 8 A.M. dual flight with Bob, the Chief Flight Instructor. When you arrived at the school, you noticed Bob drinking a beer in his car. When Bob met you in the briefing room, you could smell alcohol on his breath. You decide to cancel the flight, making up an excuse to Bob. Bob became extremely agitated and insisted that you go for the flight.

Based on this scenario, prepare a SECURITAS report to submit to the TSB.

1. Include your name, address, telephone number, profession, and experience, your involvement in the unsafe situation being reported, where else you have reported this unsafe situation or safety concern, and the identification of the aircraft and the owner of the equipment.

_____

_____

_____

_____

_____

_____

2. Describe (as appropriate) the unsafe act or safety concern in terms of how the unsafe act/condition was discovered, what happened, where, when (give date and local time) and why you think it occurred.

_____

_____

_____

_____

_____

_____

3. Give your suggestions to correct the situation.

_____

_____

_____

_____

_____

_____

CHAPTER 9

AVIATION SAFETY STATISTICS

OBJECTIVES

After you have read this chapter, you will understand:

- How aviation accident rates have decreased since the 1960s and the ramifications of the current "plateau"

- The phase of flight in which most accidents occur

- What causes most accidents

- Safety technologies such as traffic collision avoidance systems and ground proximity warning systems

Introduction

Since the 1950s, when good data regarding aviation accidents began to be collected, the worldwide accident rate has decreased substantially.

Accident Rates and Onboard Fatalities by Year
Worldwide Commercial Jet Fleet – 1959 Through 2007

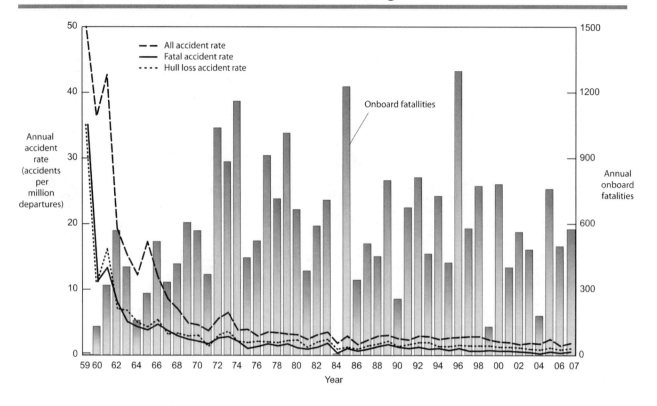

2007 STATISTICAL SUMMARY, JULY 2008

BOEING

Figure 9.1 *Accident Rates and Onboard Fatalities by Year*

The dramatic reductions in the 1960s are attributed to primarily mechanical modifications increasing the airworthiness of aircraft. In the mid-1960s, new navigation technology allowed aircraft to make precision approaches for landing which further decreased the accident rate. This technology eliminated most of the guesswork from landing and guided pilots at a safe angle of descent to the runway. Traffic collision avoidance systems (TCAS) and ground proximity warning systems (GPWS) were introduced in the early 1980s. The purpose of TCAS is to give pilots information regarding possible collision conflicts with other aircraft so that mid-air collisions can be avoided. The purpose of GPWS is to give pilots a warning when they are within near-proximity of the ground so that controlled flight into terrain (CFIT) accidents can be avoided.

It is important to note several things in this graph. The first is that in the early years of aviation, the total number of accidents was very high, but the number of fatal accidents was comparatively low. As interventions have been introduced to successfully lower the accident rate, the differential between total number of accidents and the survivability of accidents has also decreased.

The second point to consider is the number of onboard fatalities associated with accidents. Clearly, this number tends to trend upwards as larger planes are introduced over time. These numbers can be skewed heavily, based on a single large accident, and are not a good indication of the overall safety level of the aviation system. A good example of this is the Tenerife disaster in 1977. This accident involved two 747 aircrafts and resulted in 583 fatalities. Referencing Figure 9.1, it is evident that over half of the accident fatalities in that year are the result of this one accident.

The third attribute to consider, and arguably the most important, is the plateau of accident rates over the last thirty years. Despite the many safety advancements, the industry has been unable to further reduce the accident rate. These advancements include:

- Modern technology and manufacture of aircraft

- Traffic and ground avoidance systems

- New navigation technologies (such as global positioning systems [GPS])

- Human factors training (such as crew resource management [CRM])

- Safety management systems (as described in detail in Chapter 10)

At this point, some experts argue that accidents are just the cost of doing business. Certainly, any further reductions in the accident rate would be associated with tremendous costs to the industry for very small improvements. Based on this understanding, why does the aviation industry continue to invest tremendous resources in safety? This question can best be answered by the Figure 9.2.

Figure 9.2 includes the same accident rate curve that was presented in Figure 9.1. In addition, another line represents the increase in worldwide air traffic, which is expected to double by 2020! The third line presents the annual number of accidents. It is clear that even though the rate at which accidents are occurring remains constant, the total number of accidents each year will increase because more planes are in the sky and more miles are being flown. Although the accident rate remains constant, the impact that this increase may have on passenger confidence would be significant. For example, if your parents watched a news report of a major aviation accident every week, they would probably be less willing to fly. Even if they were educated about the overall accident rate remaining low and unchanged from previous years, the fear may remain. Passenger confidence in the safety of aviation directly impacts the success of the industry. After the devastating attacks of 11 September 2001, there was not necessarily an increased risk of terrorists taking over an airplane. However, it took many months for consumers to feel confident travelling again. These fears nearly bankrupt the airline industry in North America.

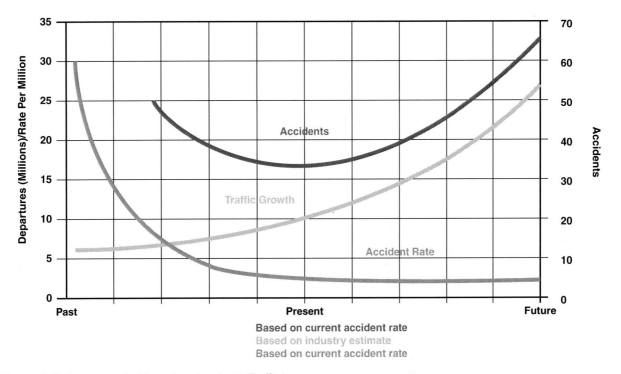

Figure 9.2 *Increase in Accidents Associated with Traffic Increase*

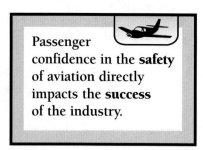

Passenger confidence in the safety of aviation directly impacts the success of the industry.

Accidents are extremely expensive to aviation companies. The cost of an accident falls under two categories: (a) costs that are covered by insurance (such as loss of aircraft, liability, and property damage), and (b) uninsured costs which can double or triple insured costs (such as the insurance deductible, fines, legal fees, increased insurance premiums, cost of investigation, loss of equipment, morale, training replacements, loss of business, damage to reputation, and rental of replacement equipment) (Flight Safety Foundation 1999).

For these reasons, it is crucially important for the aviation industry to invest resources in aviation safety. Not only do safety interventions save lives, but safety improvements also have significant economic impacts on the aviation industry.

Where Do Accidents Occur?

Within aviation, one cycle is defined as the time from takeoff to landing. A cycle consists of the takeoff, initial climb, climb, cruise, descent, initial approach, final approach, and landing phases of flight. Figure 9.3 presents accident data associated with each phase of a cycle.

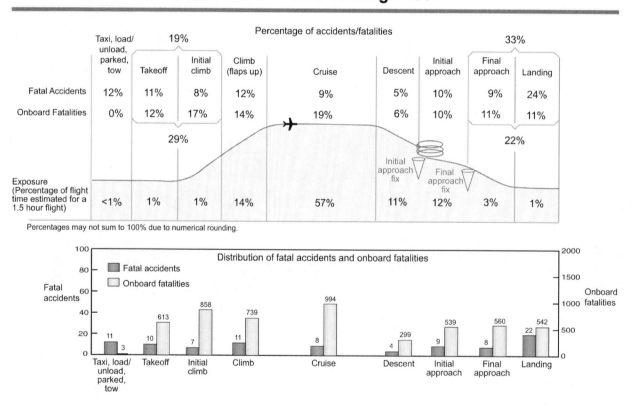

Figure 9.3 *Fatal Accidents and Onboard Fatalities by Phase of Flight*

Based on Figure 9.3, it is important to understand a few characteristics of air accidents. The exposure numbers along the bottom of the figure present the percentage of time spent in each phase of flight. What is particularly interesting is that the takeoff and initial climb phases of flight represent only 2 percent of the total flight time. However, these phases account for 19 percent of onboard fatalities. Likewise, the final approach and landing phases of flight represent 4 percent of the total flight time and 33 percent of the total fatal accidents. Overall, the takeoff, initial climb, final approach and landing represent 6 percent of the total flight time and 52 percent of the fatal accidents! Therefore, these phases have disproportionately higher accident risk than other phases of flight! For this reason, it is generally accepted that commuter airlines that fly short legs and several cycles per day will have a higher accident rate than major international airlines which fly longer, perhaps inter-oceanic, legs. This is based on the exposure to flight phases with the greatest risk—namely takeoff and landing.

Types of Accidents

When learning about aviation safety, it is helpful to have an understanding of what types of accidents are most prevalent. Figure 9.4 presents a summary of accident categories associated with the worldwide commercial fleet.

From Figure 9.4, it is clear that the majority of commercial aviation accidents are associated with LOC-I, which stands for *loss of control in-flight*, and CFIT, which stands for *controlled flight into terrain*. The less expected of these two is CFIT accidents, which basically means that an aircraft flew into the ground. Some may wonder

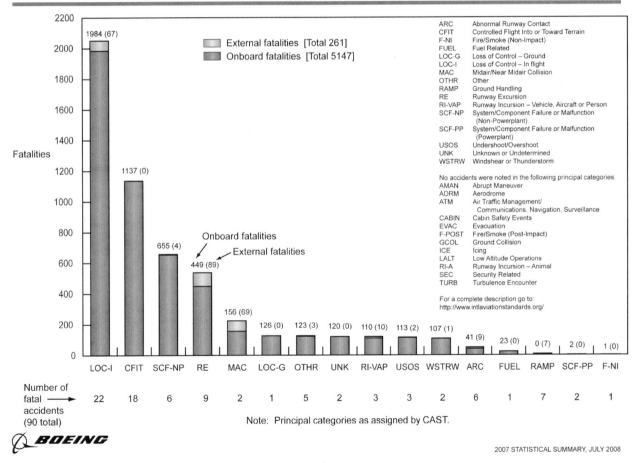

Figure 9.4 *Fatal Accidents by Category*

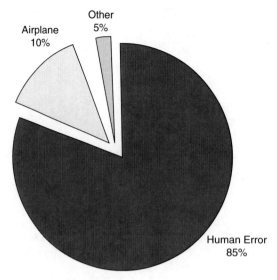

Figure 9.5 *Primary Cause of Aviation Accidents*

how this is possible in modern aviation. In almost all cases, CFIT (along with many other types of accidents) results from poor human factors leading to pilot error.

Many are surprised to learn that, by far, the majority of aviation accidents are caused by pilot error (refer to Figure 9.5). This does not mean that pilots are unsafe or bad at their jobs. Far from it! Pilots are one of the most highly-trained occupations, and they take safety very seriously. What it means is that after an aviation accident, if a mechanical failure was found to be the causal factor, it is a relatively straightforward process to identify that failure and fix all models of aircraft around the world. On the other hand, if the accident investigation reveals that human error caused the accident, it is extremely complicated to "fix" all of the pilots in the world. For this reason, accidents associated with a particular type of human error are likely to reoccur. An example of this is accidents associated with situation awareness. Situation awareness refers to a pilot's ability to maintain a mental picture of their environment, understanding what it means, and predict how it may impact the pilot in the future. CFIT accidents are examples of poor situation awareness.

Safety Technologies

There are several human factors approaches that can be implemented to reduce human error accidents. These approaches are presented in detail in Chapter 7. In addition, the aviation industry has attempted to better understand and reduce pilot error through the introduction of flight safety technologies.

Black Boxes

During the recovery process after an aviation accident, there is a lot of discussion in the media over the "black box." What they are actually referring to are two separate recorders: the Flight Data Recorder (FDR) and the Cockpit Voice Recorder (CVR). Interestingly, black boxes are actually bright orange in colour. Most commercial aircraft and some corporate and private aircraft are required to be equipped with a FDR and a CVR. These devices typically cost between $10,000 and $15,000. The memory units are located in the tail section of the aircraft (to maximize the chance of survivability in an accident). Older CVRs and FDRs record data on magnetic tape, similar to the recording tape within old audio cassettes. Newer models use solid state digital memory chips and are capable of storing greater amounts of information.

Cockpit Voice Recorder

The CVR records the voices of flight crew and other cockpit sounds. The recorder has a "cockpit area microphone" that is typically located on an overhead instrument panel between the two pilots. After an accident, an investigator will listen to the CVR to detect changes in engine noise, stall warnings, and other clicks. These noises help investigators identify when events such as system failure occur. Investigators also take great interest in the conversation between pilots, with air traffic control, and with cabin crew. This dialogue helps investigators understand what the pilots were thinking and doing before the accident occurred.

"BLACK BOX" SPECIFICATIONS

Flight Data Recorder

Time recorded—25 hour continuous

Number of parameters—18 to over 1000

Impact tolerance—3400 Gs

Fire resistance—1100 degrees Celsius for 30 minutes

Water pressure resistance—submerged to 20,000 feet

Underwater locator beacon—operates for 30 days

Cockpit Voice Recorder

Time recorded—30 minutes continuous loop, 2 hours for solid state digital units

Number of channels—4

Impact tolerance—3400 Gs

Fire resistance—1100 degrees Celsius for 30 minutes

Water pressure resistance—submerged to 20,000 feet

Underwater locator beacon—operates for 30 days

Photos.com.

Figure 9.6 *Flight Data Recorder*

NTSB Aviation, 2004.

Due to the sensitive nature of cockpit communications, CVR recordings are treated differently than other factual information gathered in an accident investigation (NTSB Aviation 2004). Unless a judge decides that the CVR recording could potentially improve aviation safety, it is unlikely that CVR data will be released publically, as it is protected by Canada's privacy act. Instead, TSB investigators will listen to the CVR privately and use the information they hear to identify causal and contributing factors in the accident.

A major limitation of older magnetic tape CVRs is that they record a maximum of thirty minutes on a continuous loop. This means that, after an accident, only the last thirty minutes of the flight have been recorded. Unfortunately, in some cases the major factor that caused the accident occurred more than thirty minutes before the impact. This results in important safety data being lost. Newer solid state recorders are capable of holding much more information and record the last two hours of the flight.

Flight Data Recorder

The FDR monitors a number of flight parameters, including time, altitude, heading, and airspeed. Newly manufactured aircraft must install FDRs that collect a minimum of eighty-eight parameters; however, some can record more than 1000. Some of these items may include auto-pilot mode or even alarms.

Based on FDR information, investigators can recreate the flight with computer animated video reconstruction This aids investigators in visualizing the flight during the moments before the accident impact with the goal of identifying causal and contributing factors (NTSB Aviation 2004).

Traffic Collision Avoidance System (TCAS)

Traffic collision avoidance systems (TCAS) were designed to reduce the possibility of midair collisions. Based on signals received from the transponders of all of the aircraft within a pilot's proximity, the TCAS presents information about where other aircraft are and identifies those that may be a collision risk. For example, the TCAS screen in the cockpit may picture another aircraft and indicate that it is at 11 o'clock (in front and slightly to the left) and 500 feet above current position and descending. If the TCAS system identifies an immediate threat, it will give the pilots instructions to avoid collisions, such as "pull up, turn right, pull up, turn right." Although pilots are now taught to follow TCAS instructions, there was confusion in the past regarding whether pilots should follow TCAS or ATC instructions.

CASE

COLLISION OVER GERMANY

On 1 July 2002, at approximately 11:43 P.M. local time, Bashkirian Airlines Flight 2937 and DHL Flight 611 collided in the skies over Germany. Both aircraft were flying at 36,000 feet and were in contact with an air traffic controller named Dane Peter Nielsen. About one minute before the collision, Mr. Nielsen instructed the Bashkirian flight to descend to 35,000 feet to avoid a conflict with the DHL flight. The crew did not respond because their TCAS was simultaneously instructing the pilots to climb to avoid the collision. The controller repeated himself and the crew acknowledged the instruction and began to descend. At almost exactly the same moment, the TCAS on the DHL aircraft determined that the Bashkirian aircraft was a collision threat and instructed the crew to descend. The DHL pilots immediately began a descent to avoid the Bashkirian aircraft. Unfortunately, the two aircraft collided at an altitude of 35,400 feet. A total of seventy-one lives were lost in the incident. On 24 February 2004, a grief-crazed man who had lost his wife and children in the accident stabbed and killed the air traffic controller who was involved in the accident (Walsh and Harding 2004).

Ground Proximity Warning System (GPWS)

Ground proximity warning systems (GPWS) were introduced into aircraft cockpits as a means to reduce the number of controlled flight into terrain (CFIT) accidents. CFIT accidents occur when qualified pilots fly a properly-functioning aircraft into mountains, ground, water, or obstacles, all the while unaware of their proximity to the surface. The GPWS alerts the cockpit crew when they are within a specific number of feet above the earth. Sometimes the term-enhanced ground proximity warning system (EGPWS) or terrain awareness and warning system (TAWS) are also used. However, all of these systems have the same purpose of alerting pilots when they are close to the ground.

GPWS was invented by a Canadian engineer named C. Donald Bateman. Bateman was born in 1932 in Saskatchewan. He received a degree in electrical engineering from the University of Saskatchewan and continues his work on EGPWS today at Honeywell (National Inventors Hall of Fame Foundation 2002).

When the GPWS identifies that the aircraft is becoming increasingly close to the surface, an aural warning will alert the pilots. This warning is typically "Terrain, Terrain" or "Pull up, pull up." Aviation accident rates associated with CFIT fell dramatically after the introduction of GPWS. This device has saved the lives of countless pilots and travelers.

Conclusion

A review of aviation safety statistics reveals that aviation is a very safe mode of transportation. However, the costs of accidents are so high that it is extremely important to continually invest resources in safety advancement. Through detailed tracking and analysis of accidents, trends begin to emerge in the accident data. Based on this knowledge, the industry is able to respond through technologies and training to enhance a pilot's ability to operate an aircraft safely.

Name _____ **Date** _____

PRACTICE ACTIVITY

You have been entrusted with $100 million to invest in aviation safety. Based on the statistics presented in this chapter, explain where you would spend the money, how the money would be divided up, and why.

_____

_____

_____

_____

_____

_____

_____

_____

_____

_____

_____

Based on the information presented in this chapter, come up with a new device that could be built into aircraft to help in accident prevention. Be creative! What would it be and how could it help?

_____

_____

_____

_____

_____

_____

_____

_____

_____

_____

_____

_____

CHAPTER **10**

SAFETY MANAGEMENT SYSTEMS

OBJECTIVES

After completing this chapter, you will:

- Understand why it is impossible to have perfect safety in aviation
- Be able to describe the strengths and weaknesses of several aviation safety perspectives
- Be familiar with the organizational perspective on safety and why is it called the *Swiss cheese model*
- Understand the components of safety management systems
- Know the difference between prescriptive vs. performance-based regulations
- Understand hazard identification and risk management processes
- Be able to describe a safety culture and why it is important in aviation organizations

Introduction

It is unreasonable and unrealistic to expect perfect safety with zero accidents in aviation. The reason for this is that making mistakes is a natural part of being human. Even philosophers in ancient Rome knew that "Errare humanum est" or "To err is human." Therefore, one can't define aviation safety as the creation of a perfect system with zero accidents because this standard is unattainable. Rather, ICAO (2008) defines safety in the following manner:

> **SAFETY IS THE STATE IN WHICH THE POSSIBILITY OF HARM TO PERSONS OR OF PROPERTY DAMAGE IS REDUCED TO, AND MAINTAINED AT OR BELOW, AN ACCEPTABLE LEVEL THROUGH A CONTINUING PROCESS OF HAZARD IDENTIFICATION AND SAFETY RISK MANAGEMENT (15).**

Therefore, safety in aviation is a systematic process that involves identifying problems and fixing them to avoid future accidents. Historically, aviation safety has been a reactive process. This means that after an accident occurs, great lengths would be taken to understand why it occurred and prevent it from happening again in the future. The problem with this approach is that an accident must occur to initiate the safety process!

This is less than desirable because lives may be lost in an accident, the public becomes increasingly nervous about air travel, the aircraft is lost, passengers or next-of-kin require financial compensation, and many other expenses occur. Therefore, a much better approach is a proactive or ultimately a predictive safety system. In these approaches, steps are made to identify hazards and risks within an aviation company and eliminate them before an accident occurs. This process is the foundation of safety management systems (SMS).

Safety Perspectives

There are several safety approaches within the aviation industry. However, SMS requires the adoption of an organizational safety perspective within the aviation industry (sometimes called the *system approach*). The organizational perspective takes the approach that the responsibility for safety does not rest solely on the shoulders of the pilot or in the structural strength of the aircraft. Rather, all elements of the organization are able to improve or diminish safety. Examples of organizational risks include management decisions regarding training budgets and types of aircraft, maintenance shortcuts, and chief pilot pressure to get the job done in less-than-perfect weather conditions. All these factors diminish the level of safety in an organization, although they probably won't lead to an accident immediately.

To understand the organizational approach to safety, it is helpful to review other safety perspectives for comparison. All of these approaches are used by aviation professionals in one manner or another to enhance safety. However, there are also shortcomings associated with each approach.

Mechanical

Early aviation safety focused primarily on the aircraft as the source of accidents. After an accident occurred, the investigation process would identify risky or faulty components and replace these components on other aircraft models. This approach would involve a thorough investigation of the aircraft itself. Although structural components fail occasionally, this approach cannot identify all potential sources of error. For example, an accident is sometimes the result of an encounter with a wind shear. This condition would not be identified by investigators who focus solely on the mechanical components of the aircraft.

Environmental

Aircraft must avoid dangerous weather conditions, including thunderstorms and icing. If pilots are unaware or unprepared or their aircraft are unequipped for dangerous weather conditions, accidents may result. This approach has led to the development of effective onboard weather radar and anti-icing systems. However, an environmental approach to accidents does not account for controlled flight into terrain on a bright and sunny day.

Social

This approach attempts to understand safety as a function of the social environment. With this understanding, an organization would examine the social dynamics within flight crews to determine when risks and mistakes occur. Based on this determination, crew training would be offered to teach pilots how to avoid mistakes. This approach has some advantages, as it has resulted in crew resource management (CRM) training, which is currently required for most airline and military aviators. However, this approach also has some shortcomings, as all accidents cannot be attributed to crew communication errors. For example, sometimes the pilot becomes medically incapacitated.

Medical

This approach describes accidents as the result of medical deficiencies in the pilot's physiology. Accidents may result from pilot fatigue, medication complications, heart-attacks, and other physical manifestations of illness. This perspective is the reason all pilots must maintain pilot medical certificates. However, this approach has limitations, as it does not account for other types of error. For example, if the man–machine interface is poorly designed, the pilot may program the flight management system incorrectly. It is very difficult, if not impossible, to link this error to a medical condition.

Psychological

This perspective describes accidents as the result of limitations of human thinking. For example, when pilots experience a controlled flight into terrain, it often occurs without the pilot being aware of the situation or surroundings. This lack of situation awareness is linked to the limitations of human memory—a cognitive process. To improve safety, new avionics can be integrated that provide the pilot with an enhanced display of external terrain. This display would reduce the mental workload on the pilot so that he or she can pay attention to other aspects of the flight. However, this approach does not account for other types of accidents, such as those associated with a mechanical breakdown.

Ergonomic

A significant advancement in safety science was the Ergonomic perspective which was promoted by ICAO in the early 1990s. This approach is slightly more complicated than the rest, viewing safety as the result of the interaction among humans, machines, and the workplace. The best known example of this perspective is the SHELL model. The SHELL model describes accidents as a result of the interaction between the pilot and their software, hardware, the environment, and liveware (Hawkins 1987). *Software* refers to the operational context which includes standard operating procedures, checklists, symbols, and aviation regulations. *Hardware* refers to the pilot's physical environment including the cockpit, control yoke, machinery, seats, and maintenance of the aircraft. *Environment* refers to both the social environment in the workplace and the external environment (such as heat, vibration, and lighting levels). The *liveware* refers to all people in the system, including ATC, dispatch, other pilots, cabin crew, and passengers. Through this model it is possible to see how each factor in the SHELL model could possibly contribute to or diminish the level of safety within the aviation system.

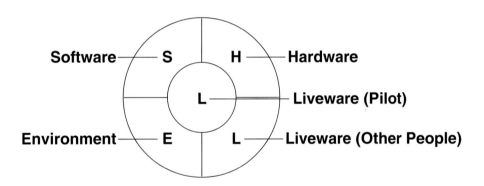

The Organizational Perspective

All of these approaches contribute to our understanding of how safety and accidents occur. However, only the **organizational** perspective considers all of these strategies and places them within the context of an organization. In the 1990s, a book by James Reason revolutionized how the aviation industry understood accidents. Within this book, accidents were described as being the result of risks and unsafe conditions that are present within every organization. Reason (1990) defines the productive activity of aviation to be safe and efficient flights. Yet, before a flight occurs, preconditions are established within an organization. Examples of preconditions include the maintenance of equipment, level of training of the workforce, and operations schedule. From this perspective, one can appreciate that before a pilot reaches an aircraft, he or she is dependent on how all other elements of the organization have contributed to the safety of the flight. Therefore, organizations need to look beyond the cockpit to identify all contributing factors that lead to an accident!

AIR ONTARIO 1363: ONE OF THE FIRST ORGANIZATIONAL ACCIDENT INVESTIGATIONS

By Jason Fogg

The crash of Air Ontario Flight 1363 on 10 March 1989 in Dryden was not just another aircraft icing accident in Northern Canada. After the Canadian Aviation Safety Board's (CASB) poor handling of Canada's worst accident—Arrow Air Flight 1285—just four years earlier, the same story was unfolding again. The aviation industry was in dire need of immediate change to a more organizational focus. Through the Honourable Virgil P. Moshansky's Commission of Inquiry into the Dryden Accident, this process was finally started.

BACKGROUND

The aviation industry in Canada had undergone many changes leading up to the Air Ontario crash. Most importantly, Canada had finally followed the United States and deregulated their airline industry in 1984. This allowed carriers to compete on routes anywhere in Canada. By the late 1980s, hundreds of new airlines were popping up in North America. Next, the Canadian government slowly let go of its flagship airline, Air Canada, and by 1989 all of its shares were sold to private investors. This process gave Air Canada and its feeder airlines freedom to carry out the business as they saw fit, but also made them financially vulnerable without the protection from Ottawa.

The accident investigation system in Canada was coming under fire in the late 1980s. In 1985, Canada suffered its worst aviation accident with the crash of Arrow Air Flight 1285. Intense public controversy followed the CASB handling of this horrific crash that killed 248 U.S. military personnel and eight crew members. The CASB was scrutinized for its search for a probable cause while missing out on several other preconditions that may have contributed to the accident. When Air Ontario Flight 1363 crashed just four years later, the Minister of Transport halted the CASB investigation and initiated a Commission of Inquiry led by the Honourable Virgil P. Moshansky.

QUICK FACTS ABOUT THE ACCIDENT

*   When: 10 March 1989

*   Where: Thunder Bay to Winnipeg, via Dryden

*   What: Fokker F-28 unable to take off due to icing

*   Who: 21/65 passengers and 3/4 crew killed

*   Why: More than just aircraft icing

MOSHANSKY'S ORGANIZATIONAL APPROACH

Moshansky called his style of investigation a "system-analysis" approach and focused on five key areas: the aircraft, crew, operational environment, air carrier, and regulator. His analysis spanned more than three years and included 166 witnesses, 34,000 pages of testimony, and 177,000 pages of exhibits. Mr. David Wightman, assistant deputy minister of transport said that the inquiry was "the most in-depth look at the operations of Transport Canada . . . and the regulatory side of it specifically, that we've ever had" (Moshansky 1992).

Aircraft ↔ Crew ↔ Operational Environment ↔ Air Carrier ↔ Regulator

FINDINGS AND RECOMMENDATIONS

Whereas previous investigators may have simply blamed this accident on pilot error, Moshansky made countless findings during his inquiry into the Dryden accident in each of the five areas. Some of the most important of those findings were:

- The F-28 was new to the Air Ontario fleet; both the captain and first officer had less than 100 hours experience on this aircraft type.

- Air Ontario's flight safety officer had just quit, with no one filling this role at the time of the accident.

- Air Ontario had two previous emergency landings due to icing, but neither was investigated by Transport Canada.

- There was no functioning Auxiliary Power Unit (APU) in Dryden, meaning the aircraft engine's would need to remain running during the station stop.

- The aircraft was allowed to "hot refuel" with passengers on board, but there were no procedures in place for de-icing with the engines running.

From these findings, Moshansky made 191 recommendations to the Minister of Transport in the interest of aviation safety. A few of the most important to consider were:

- Prohibit refueling of aircraft with passengers on board.

- Immediate upgrades to de-icing equipment were needed at major airports (i.e., CYYZ).

- Research and development was needed into critical surface contamination.

- Many critical changes to the investigative process were needed; Moshansky suggested a more organizational approach like the one he used in his inquiry.

CHANGES TO AVIATION

The Minister of Transport took the inquiry very seriously and implemented many of Moshansky's recommendations directly into the Canadian aviation industry. These included a number of documents relating to icing, such as the Transport Canada publication, *When in Doubt . . . Small and Large Aircraft— Aircraft Critical Surface Contamination Training for Air Crew and Ground Crew* (TP10643). However, the most important changes to the aviation industry had a more widespread application. Shortly after the Dryden Accident, Transport Canada initiated a program whereby air carriers would be held accountable for their own flight safety program. The development of this process led to what is now commonly known as *Safety Management Systems* (SMS).

Organizational influences on safety are described by Reason (1990) as existing in four layers: (1) organizational influences, (2) unsafe supervision, (3) preconditions for unsafe acts, and (4) unsafe acts. These four layers are often pictured as slices of Swiss cheese. The holes in the Swiss cheese are latent conditions. Latent conditions are risk factors that do not lead to an accident immediately; rather they lie dormant within an organization, often for years. Examples of latent conditions include unsafe maintenance practices, poor training due to management cutbacks, and a policy of hiring pilots with minimal experience to save salary costs. Clearly, all of these factors reduce the level of safety within layers of the organization. However, none of these would lead to an accident immediately. It is only when holes in all layers of Swiss cheese align that an opportunity for an accident occurs.

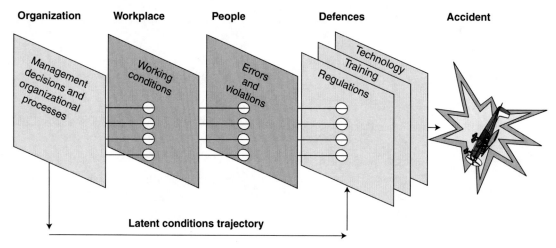

ICAO 2008; Reason 1990.

Figure 10.1 *Reason's model, often referred to as the "swiss cheese" model.*

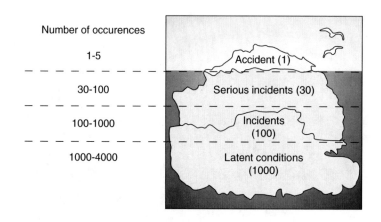

ICAO 2008.

Figure 10.2 *The key to accident prevention is identifying unsafe latent conditions and eliminating them before an accident occurs.*

This perspective allows companies to adopt a proactive/predictive approach to safety. Rather than waiting for an accident to occur before safety risks can be identified, now organizations can attempt to identify the "holes" in the cheese before these latent conditions line up to cause an accident. Latent conditions that are present within a company will manifest as minor incidents before materializing as a major accident. Therefore, through monitoring minor incidents and conducting active hazard identification and risk assessment processes, a company can identify latent conditions pre-emptively.

Safety Management Systems

The process of identifying latent conditions before an accident occurs and eliminating them is an important component of safety management systems (SMS). SMS is defined as a "formalized framework for integrating safety into the daily operations of a transportation enterprise" (Transport Canada 2007, 15). The goal of SMS is to proactively manage risks in an effort to prevent accidents from occurring. This is accomplished through the following four approaches:

- Defining clear accountability and responsibility

- Setting performance goals and actively pursuing them

- Managing safety risks systematically and proactively, including continuous improvement and learning
- Monitoring and evaluating performance towards goals (15)

Within the aviation industry in Canada, SMS is being implemented into five major sectors:

1. Aeronautical product design and manufacturing
2. Air operations
3. Aircraft maintenance
4. Airport operations
5. Air navigation system

The three foundational characteristics of SMS are that it be systematic, proactive, and explicit.

Systematic refers to safety activities occurring based on a plan that is applied consistently throughout a company. In addition, a long-range plan is developed, approved, implemented, and operated on a continual day-to-day basis. Therefore, improvements from SMS will not occur immediately, but gradually over time. The main focus of SMS is to identify hazards within the operation before an occurrence materializes (ICAO 2008).

SMS is proactive as it uses hazard identification and risk management to identify problems before an accident or incident occurs. Instead of reacting after an accident, SMS utilizes strategic planning to control the level of organizational risk. To be effective, proactive SMS requires constant monitoring of operations to collect safety data on hazards. This allows for decisions to be made based on collected safety data rather than opinion.

Finally, SMS is explicit as all safety activities are documented, visible, and defensible. Safety training and activities of all personnel are formally recorded in official documentation that is available to anyone. The requirement of SMS to be explicit supports the importance of a good safety library within an organization.

Key Components of SMS

Safety management can be broken down into eight key components, as described by ICAO (2008).

1. Commitment of senior management to safety management: In order to have the resources available to manage safety, senior management must be in full support of SMS in order to produce successful results. Quite simply, SMS cannot exist without the financial resources required to support the process.

2. Safety reporting: A continual stream of data is required to identify hazards and risks within an organization. Therefore, voluntary and confidential safety reporting should be encouraged among operational employees.

3. Continual monitoring: Systems that collect safety information during day-to-day operations must be continually analyzed, extracted, and interpreted to identify risks. These findings must be distributed to the operational personnel who work with these systems each day.

4. Safety occurrence investigation: Reactive investigations are important to identify why accidents happened. However, it is important that these investigations do not focus on identifying a "guilty" party to assign blame, as this will not reduce the number of organizational accidents.

5. Sharing safety lessons: The results of SMS processes should be widely distributed throughout the aviation industry so that others can avoid repeating the same mistakes.

6. Integration of safety into operational personnel training programs: In the past, a common perspective in the industry was that skilled operators are safety experts on their task. However, it is now evident that this is not an accurate approach. Rather, operational personnel should have dedicated training on the foundation of safety management.

7. Utilization of Standard Operating Procedures (SOPs) throughout the organization: SOPs are standardized step-by-step instructions for operators (such as pilots, maintenance staff, and air traffic controllers) that detail the upper management's desired way of completing a specific procedure or

manoeuvre. In this manner, management is capable of controlling the actions of a large number of operational personnel in the interest of safety. When realistic well-written SOPs are utilized within an organization, they are powerful safety tools.

8. Continuous improvement of safety: Finally, it is important to appreciate that safety issues cannot be solved. The entire organization must approach safety as an important component of day-to-day operations that can only succeed through continual improvements.

Previously, Transport Canada managed safety by stepping in at the operational level. Under SMS, Transport Canada interacts with companies at the organizational level to ensure that an appropriate safety management process is in place and daily operations are compliant. If an operator's SMS program is found to be running poorly, Transport Canada will intervene. Therefore, Transport Canada maintains the ability to utilize traditional compliance inspection and auditing activities.

Prescriptive vs. Performance-based Regulations

As of 2010, Safety Management Systems are an international requirement. However, Canada has been a world leader in the early implementation of SMS. The implementation of SMS is evidence of a shift from prescriptive to performance-based regulations.

Old prescriptive regulations were designed as administrative controls. To ensure safety within an organization, a strict regulatory framework was established that included regular safety inspections and audits by Transport Canada. Under prescriptive regulations, to maintain safety, aviation companies were required to comply with the regulations. However, just because an organization complies with prescriptive regulations, there is no guarantee of safety!

> "Transport Canada's policy is for industry to be accountable for systematically and proactively managing risks and threats within their transportation activities."
>
> Transport Canada 2007, 9.

Modern performance-based regulations represent a major shift in safety policy. These regulations are designed as safety risk controls, rather than as administrative requirements. Performance-based regulations allow companies to comply with regulations dynamically, based on their unique identification and prioritization of safety risks (ICAO 2008). Therefore, the manner in which each company complies with performance-based regulations will vary slightly depending on the areas of risk they have identified within their specific organization. Performance-based regulations are viewed as "smart regulations." "Transport Canada's policy is for industry to be accountable for systematically and proactively managing risks and threats within their transportation activities" (Transport Canada 2007, 9).

With Transport Canada's shift away from being an owner-operator of airports and air traffic control services, to the primary role of providing regulatory oversight, a greater percentage of safety responsibility lies with the industry. Transport Canada's performance-based regulations are a part of this evolution.

TRANSPORT CANADA'S PERFORMANCE-BASED SMS REGULATIONS REQUIRE THE FOLLOWING:

- A SAFETY POLICY

- A PROCESS FOR SETTING GOALS FOR THE IMPROVEMENT OF AVIATION SAFETY AND FOR MEASURING THE ATTAINMENT OF THOSE GOALS

- A PROCESS FOR IDENTIFYING HAZARDS TO AVIATION SAFETY AND FOR EVALUATING AND MANAGING THE ASSOCIATED RISKS

- A PROCESS FOR ENSURING THAT PERSONNEL ARE TRAINED AND COMPETENT TO PERFORM THEIR DUTIES

- **A PROCESS FOR THE INTERNAL REPORTING AND ANALYZING OF HAZARDS, INCIDENTS, AND ACCIDENTS AND FOR TAKING CORRECTIVE ACTIONS TO PREVENT THEIR RECURRENCE**

- **A DOCUMENT CONTAINING ALL SAFETY MANAGEMENT SYSTEM PROCESSES AND A PROCESS FOR MAKING PERSONNEL AWARE OF THEIR RESPONSIBILITIES WITH RESPECT TO THEM**

- **A PROCESS FOR CONDUCTING PERIODIC REVIEWS OR AUDITS OF THE SAFETY MANAGEMENT SYSTEM AND REVIEWS OR AUDITS FOR CAUSE OF THE SAFETY MANAGEMENT SYSTEM**

- **ANY ADDITIONAL REQUIREMENTS FOR THE SAFETY MANAGEMENT SYSTEM THAT ARE PRESCRIBED UNDER THESE REGULATIONS**

In addition to improving the level of safety within an organization, there are many other benefits associated with the implementation of SMS. Some of these benefits include:

- Direct and indirect savings resulting from the prevention of accidents (including decreased insurance premiums). Several aviation companies, including Skyservice Airlines, Air Transat, Conair, Harbour Air, and Moncton Flight College have publically demonstrated how SMS has improved their financial performance (Transport Canada 2007).

- Aviation companies will be more competitive in the global market.

- Due diligence is easily demonstrated after the fact if an accident occurs, because of the detailed documentation completed within SMS.

- Improved relationships, morale, and productivity with employees

- Increased knowledge and management of safety by operational personnel who are able to make the greatest positive impact

- Improved relationships with industry partners through safety collaborations and the sharing of information

- Increased collaboration between industry stakeholders on safety issues to reduce risk, particularly in emergency preparedness activities (Transport Canada 2007)

Hazard Identification and Risk Management

You can't manage what you can't measure!

Hazard identification and risk management are two approaches within SMS that are utilized to improve safety. These concepts were not developed specifically for SMS, as they have a history reaching back more than forty years (ICAO 2008). The difference between hazards and risks are often confused. A hazard is something with the potential to cause harm. Risk, on the other hand, builds on the understanding of hazards to include a measure of how severe and how likely the consequences are.

The following descriptions provide an overview of how hazard identification and risk management processes are utilized within SMS.

Hazard Analysis

Hazard analyses consist of four steps:

1. Understanding
2. Identification
3. Analysis
4. Documentation (ICAO 2008)

Understanding Hazards

A hazard is defined as "a condition or an object with the *potential* of causing injuries to personnel, damage to equipment or structures, loss of material, or reduction of ability to perform a prescribed function" (ICAO 2008, 57). Therefore, hazards should not be thought of as bad or dangerous. Rather, they are a natural part of operations within an organization.

A separate component of a hazard is its consequence. A consequence is defined as "the potential outcome . . . of a hazard" (ICAO 2008, 58). The consequence is distinct from the hazard itself, although this point is often confused. For example, a foreign object lying on a runway is a **hazard,** while damage resulting from the foreign object being sucked into an aircraft engine is a **consequence**. A hazard often exists without a consequence.

Therefore, hazards exist in the present, while consequences exist in the future and can be predicted.

Identifying Hazards

Within an aviation organization's SMS, hazards can be grouped into two categories:

1. Natural hazards—the result of environmental factors including weather, geography, and wildlife

2. Technical hazards—the result of aircraft system or component failure, ground equipment or tool failure, or facility malfunctions (ICAO 2008)

Within these two categories, there is a wide range in the types and sources of hazards within an organization. To identify hazards as part of safety management, several data sources can be utilized, including flight data analysis, safety surveys, voluntary safety reports, feedback from training, and accident reports.

Analyzing Hazards

Once a hazard has been identified, it must be analyzed. Hazard analysis is a three step process:

1. Identification of the generic hazard

2. Deconstruction of the generic hazard into specific hazards or components

3. Identification of potential consequences of specific hazards

For example, in an effort to reduce salary costs, a small commuter airline that operated in a mountainous area has decided to grant retirement packages to several senior pilots and hire young inexperienced pilots as replacements. The following is an example of a hazard identification process based on this example:

1. Generic hazard

    a. Hiring low-time pilots

2. Specific hazards

    a. Pilot forgetting standard operation procedures (SOPs)

    b. Pilot losing situation awareness

    c. Etc.

3. Specific consequence

    a. Aircraft making unsafe approach to land, resulting in hard landing

    b. Aircraft experiencing controlled flight into terrain (CFIT)

    c. Etc.

Documenting Hazards

The final step in the hazard identification process is documentation. The documentation of hazards within an organization is an important tool that leads to the establishment of a safety library. A good safety library includes the following characteristics:

- Definition of terms

- Understanding of terms

- Safety information validity

- Amount and type of reporting the organization expects

- Measurement and management of safety information (ICAO 2008)

Risk Analysis

Once the hazard identification process is complete, the risk analysis process begins. Risk analysis goes beyond the findings of hazard identification to determine the *severity* of consequences. This is the purpose of risk analysis—to quantify (assign a number to) the consequences of a hazardous situation. This is an important step as it allows decision makers to effectively target resources towards strategies that will have the greatest impact on organizational safety.

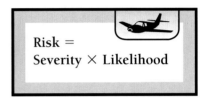

Risk =
Severity × Likelihood

Risk can be defined as the expected injuries or losses associated with an event. It is calculated as Risk =Severity × Likelihood (R = S × L). This calculation forms the basis for a risk matrix. A risk matrix is a table that plots severity along one axis and likelihood along the other.

Risk ratings between 0 and 4 are considered "low." Items with low risk are considered acceptable, although they should be monitored for continual improvement. Risk ratings between 5 and 14 are considered "moderate." Moderately risky items should be maintained as-low-as-reasonably-practicable (ALARP). In this category, cost-benefit analyses are required to determine if intervention is justified. Risk ratings between 15 and 25 are considered "severe" and are intolerable within an organization. Immediate intervention is required in this situation to reduce the level of risk.

For example, pilot fatigue on long-haul flights may have only a "moderate" level of severity, as it leads only to small mistakes in the cockpit and a second crew member is present to catch these slips. However, the company's flight schedule consists of mostly overnight delivery flights. Therefore, the possibility of this hazard would be "likely." In this situation, the overall level of risk is severe and intolerable within an organization.

A risk matrix is used as a tool to determine whether an event is associated with low, medium, or high risk. A risk matrix is helpful in quantifying the amount of risk in a situation. Assigning a number to the amount of risk gives decision makers solid information about how to respond to the situation.

		LIKELIHOOD				
		Remote	Rare	Unlikely	Possible	Likely
		1	2	3	4	5
Zero Effect	0	0	0	0	0	0
Small Effect	1	1	2	3	4	5
Minor	2	2	4	6	8	10
Moderate	3	3	6	9	12	15
Severe	4	4	8	12	16	20
Catastrophic	5	5	10	15	20	25

SEVERITY

Table 10.1 *Risk Matrix*

Risk Management Process

Transport Canada's risk management process (2004) includes the following six steps. It is important to note that thorough documentation is an important component of all six steps.

Step 1—Initiate process.   Before a risk management process can begin, a thorough evaluation of the situation or activity must be conducted. This is accomplished through:

- Describe the situation: All background information about the situation, including associated issues, must be documented.

- Identify hazards associated with the situation: Based on the definition and description provided earlier in this chapter, a listing of hazards associated with the situation must be identified.

- Identify a risk management team of subject matter experts: Team members are chosen based on how closely their expertise relates to the situation being evaluated. An important aspect of identifying the team is establishing who has authority over the group.

- Identify all internal, external, or public stakeholders: Internal stakeholders are persons or groups who will be impacted by the assessment within the same organization. External stakeholders are persons from outside corporations, federal departments, or groups from the aeronautical industry. Public stakeholders are persons from private corporations or from the general public.

- Determine if stakeholders must be consulted to analyze the risk: A determination will be made as to whether the team will require information from stakeholders.

Step 2—Perform preliminary analysis and estimate risk.

- Identify the specific components of the hazard and the risks associated with each.

- Determine the exposure interval: The exposure interval refers to the period of time within which the hazard may occur (for example, the life of an aircraft or the contract of a specific individual).

- Develop hazard scenarios: For each hazard identified, sequences of events are created to describe what could happen in the particular situation.

- Identify the consequences for each hazard scenario.

- Determine the level of risk using a risk matrix (Risk = Severity x Likelihood).

Step 3—Evaluate the risk activity.

- Identify the risk activity.

- Perform a cost and benefit analysis—this is accomplished by listing the financial costs and benefits of the risk activity (cost vs. revenue). In addition, the nonfinancial costs and benefits are considered (such as quality of life or public perception).

- Summarize the costs and benefits in terms of whether or not the activity makes sense.

Step 4—Control risk.

- Brainstorm the possible methods for controlling the risks associated with hazard scenarios. Risk is mitigated through four strategies outlined by the TEAM acronym: Transfer risk, Eliminate risk, Accept risk, or Mitigate risk.

- Assess which risk control option is most desirable.

- Determine what residual risks are associated with the chosen option.

- Select the final risk control option.

Step 5—Take action.

- Develop a risk control implementation plan: Information such as the start date, due date, person accountable, milestone/deliverable, and follow-up method should be included.

Step 6—Monitor impact and follow up.

- Identify the activities that must be monitored to provide an indication of whether the process is working.

- Identify when these activities will be monitored.

- Determine how these activities will be monitored.

- Assess how effective the risk control measures are.

- Evaluate the effectiveness of the risk analysis process.

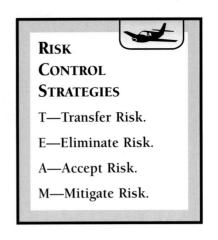

RISK
CONTROL
STRATEGIES

T—Transfer Risk.

E—Eliminate Risk.

A—Accept Risk.

M—Mitigate Risk.

Cost-Benefit Analysis

In Step 3, an important component of risk management is conducted; the cost-benefit analysis. This step is crucial to ensure that the resources invested in safety activities do not end up bankrupting the company.

For example, it is stereotypical for an aviation company to have slogans proclaiming that safety comes first. However, the reality is that most organizations are not in the business of safety. They are in the business of aviation, and they will not survive if they do not generate profits. Therefore, although safety is a very important aspect of business, it is one of several organizational objectives. Often this creates a dilemma for management as they can choose to invest resources into production activities (such as purchasing new aircraft, establishing new routes, or fostering new markets) or into protection activities (including maximizing training, incorporating shorter intervals between maintenance inspections, or hiring only senior experienced pilots). Production and protection activities should be balanced, as too many resources invested in one often results in sacrifices in the other. Investing too much into protection activities may lead to bankruptcy, while too much into production may lead to an accident. To balance an organization's production and protection goals, analysis is required.

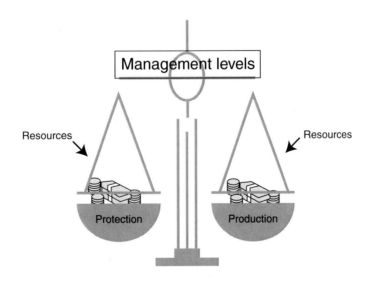

ICAO 2008.

Figure 10.3 *The production vs. protection balance.*

SAFETY METHODS

Differences Among Reactive, Proactive, and Predictive

In order to deal with the errors and risks that are present within every organization, reactive, proactive, or predictive methods are used. The purpose of these methods is to capture data that can be analyzed to provide direction as to how an organization can avoid unsafe situations (ICAO 2008). These methods exist on a continuum from least efficient (reactive), moderately efficient (proactive), to most efficient (predictive). It is crucial to understand that each of the following methods captures a different aspect of safety risk. Therefore, complete and mature safety management systems should incorporate all three of the following safety methods.

Reactive

Reactive methods respond to an event that has already occurred. As mentioned previously, aviation safety has historically been a reactive process, waiting for an accident to occur before taking any action. However, there is a place for reactive methods within modern SMS. For example, sometimes unusual events occur that are entirely unpredictable. If SMS focused exclusively on predicting accidents before they occur, this unusual event would receive minimal attention. Reactive investigations play an important role in SMS. However, the effectiveness of reactive measures is dependent on whether the investigation explores beyond the cause of the accident to identify the factors that contributed to the accident and other organizational safety risks. Examples of reactive measures include accident and incident investigations (ICAO 2008).

Considering Reason's Swiss cheese model, it is apparent that reactive methods identify latent conditions in only the final layer of protection before an accident occurs. Therefore, although a reactive approach is an important aspect of SMS, data from this method alone is insufficient for effective management of safety.

Proactive

A proactive method involves actively trying to identify safety risks within an organization before an accident or serious incident occurs. Typically, a minor safety event with little or no damage triggers a proactive process. This method is based on the idea that catastrophic events can be minimized by:

1. Identifying risks before a serious failure occurs, and

2. Working to minimize these risks.

Examples of proactive methods include mandatory and voluntary reporting systems, safety audits and safety surveys (ICAO 2008). Compared to reactive methods, proactive methods identify additional latent conditions ("holes in the cheese") further up the chain. Overall, proactive methods produce a very efficient safety strategy. However, by the time latent conditions are identified by proactive methods, they have had the opportunity to cause some form of damage or risk to the company. Therefore, reactive methods should be used to develop redundant safety methods that are capable of catching accidents before they occur.

Predictive

Predictive methods are not based on the occurrence of unsafe situations. Predictive methods analyze real-world normal operations within a company with the goal of identifying future problems. Predictive processes occur on a continual basis, taking the perspective that the safest system should actively look for trouble rather than wait for something bad to happen. Predictive methods are largely based on gathering enormous amounts of data throughout an organization, analyzing the data (often by comparison to the findings of reactive and proactive methods), and identifying trends. Examples of predictive methods include confidential reporting systems, monitoring of normal operations, and analysis of flight data (ICAO 2008).

Predictive methods are the most efficient method of identifying latent conditions at high levels in the system before they materialize into an accident. This approach identifies organizational hazards at the point of origin before they lead to a dangerous situation. Therefore, the safety nets developed from predictive methods can potentially block hazards entirely.

Safety Culture

Through the implementation of SMS within the aviation industry in Canada, Transport Canada hopes to establish a strong safety culture. A safety culture is the shared set of beliefs, attitudes, and values in regards to safety among employees. The development of a safety culture is an important aspect of SMS, and SMS represents a cultural shift in aviation. Previously, the culture of aviation often reinforced that safety was management or Transport Canada's responsibility. A safety culture fosters the attitude that safety is a responsibility that is shared equally among all employees in an organization, from senior management to operational personnel.

Within a safety culture:

- Management is accountable and responsible for safety.

- Everyone in the organization takes an active role in safety.

- Activities are risk-based.

- Organizations focus on systemic causes of accidents and incidents.

- Investment is made in proactive activities.

- Performance is measured.

- The system is audited.

- Safety, ongoing learning, and continual improvement are standard in the aviation industry, despite high workload, times of financial crises, or labour pressure (Transport Canada 2007, 15).

Some of the activities that demonstrate a strong safety culture include the involvement of employees in safety management, the number of hazards and unsafe events that are reported (as this is an indication that employees feel comfortable submitting safety reports), the sharing of safety information with other companies, active hazard analysis, consideration of safety impacts of every day-to-day decision, and the achievement of continual safety improvement (Transport Canada 2007).

Overall, a safety culture is something that must be fully integrated into the daily operations of a business. It is not something that begins and ends at monthly safety meetings. Rather, safety should become an important component of the way business is done.

Conclusion

After reviewing the various approaches to aviation safety, it is apparent that the organizational perspective (also called the *Swiss cheese model*) is the most advantageous. This approach allows for the identification of hazards and risks before they materialize into a costly accident or serious incident. This proactive/predictive approach to aviation safety involves continual hazard identification and risk management processes.

Transport Canada has been a world leader in the implementation of SMS. This leadership has shifted the traditional regulatory culture of prescriptive to performance-based "smart" regulations. Performance-based regulations allow organizations to comply by promoting safety in a manner that is specific to their operations. In addition, these regulations are fostering a safety culture throughout the aviation industry across Canada.

PRACTICE ACTIVITY

As discussed in this chapter, it is unreasonable to expect perfect safety in aviation. Based on this understanding, complete a hazard identification process for the following fictional scenario:

A local flight school is implementing SMS for the first time. A young pilot who has just completed the flight instructor rating is hired and given the title of safety officer. In observing the organization, several issues come to mind. First, the chief pilot has a very intimidating personality, which causes many staff members to be nervous about reporting safety issues. Second, because of a lot of hiring at the airline level, very few senior instructors are on staff. The school is recruiting new instructors, but they all possess minimal experience. Finally, the flight school recently replaced its Cessna 172 aircraft fleet with Diamond DA-20s. Very few staff members have experience with this type of aircraft.

Identify Hazards

What sources of information would you review to identify hazards (safety surveys, flight data analysis, etc.)? Describe how you would gather this information. Identify the natural and technical hazards in this scenario:

_____

_____

_____

_____

_____

_____

Analyze Hazards

Once you have identified some hazards, analyze them in the following three-step process (an example is included earlier in this chapter):

1. Identify the generic hazard.

_____

_____

_____

_____

_____

_____

_____

_____

_____

_____

_____

2. Break apart the generic hazard to identify specific hazards.

_____

_____

_____

_____

_____

_____

_____

_____

_____

_____

_____

3. Identify potential consequences of the specific hazards.

_____

_____

_____

_____

_____

_____

_____

_____

_____

_____

_____

CHAPTER **11**

THE FUTURE OF AVIATION

OBJECTIVES

After completing this chapter, you will be familiar with several student opinions on which issues and technologies will have the greatest impact on the future of the aviation industry, including:

- Synthetic vision

- Airbus 380 Super Jumbo aircraft

- Blended wing body aircraft

- New and emerging markets

- Automation

Synthetic and Enhanced Vision Systems

Reducing Human Error in Low Visibility Aircraft Operations

By Joshua Norman

Introduction

Over the last ten years, the accident rate for U.S. and Canadian air operators has been relatively constant at less than one accident per one million departures (Boeing 2008). During the same time period, Controlled Flight into Terrain (CFIT) has been responsible for 1137 of 5147 onboard fatalities, second only to in-flight loss of control (Boeing 2008). As air traffic density continues to increase while the accident rate per one million departures remains constant, synthetic and enhanced vision systems offer the best chance of reducing aviation incidents and accidents over the next ten years. In turn, this will reduce the high number of deaths associated with CFIT events. This will be accomplished primarily by increasing pilot situational awareness in low-visibility situations. As this technology is further developed for use in the fixed-wing environment, headway will continue to be made in adapting it to the specific needs of helicopter operations.

Although the basic idea of a Synthetic Vision System (SVS) has been around since the 1950s, it was not until 1994 that NASA, the FAA and private industry began to seriously explore the possibility of augmenting Primary Flight Displays (PFDs) and Heads Up Displays (HUDs) with terrain data from an onboard database (Pope 2006). The goal of this fusion of data is to provide VFR-like situational awareness in IFR conditions (Hemm and Houser 2001). In 1999, NASA's Aviation Safety Program began testing SVS prototypes onboard aircraft previously owned by the U.S. Air Force (NASA 2000). Since then, additional research has been completed by NASA on various commercial aircraft, including Boeing 757s and Gulfstream G-Vs. Research is still being conducted directly by NASA in hopes of reducing fatal accident rates by 80 percent from 2001 levels, before the end of 2011 (Scott 2004). It is important to note that in contrast to SVS, Enhanced Vision Systems (EVS) use sensors to supply live data to pilots through both PFDs and HUDs (Arthur, Kramer, and Bailey 2005).

Although SVSs are still being researched and improved, they have shown so much promise that many aircraft manufacturers are beginning to integrate them into current product offerings. Currently, Diamond Aircraft offers an optional SVS integrated into the Garmin G1000 on their DA-40, while private-jet manufacturers Dassault, Learjet, Gulfstream, and Cessna are beginning to incorporate them into their avionics packages (Watson 2008). It is important to note that both Airbus and Boeing do not offer SVS in any of their avionics packages at this time, and appear hesitant to adopt it. For example, Airbus's Manager for Navigation Systems, Mike Abbott, makes the argument that their enhanced ground proximity awareness system and other offerings currently used on Airbus aircraft do more than enough to facilitate safe flight within a tight air corridor (Associated Press 2008). Current research does not appear to support this claim.

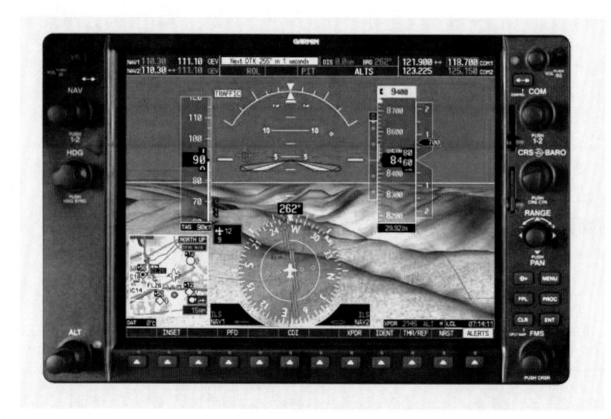

Figure 11.1

Role of Synthetic Vision in Fixed Wing Air Operations

Integrating SVSs into Current Technology

Synthetic Vision Systems have been identified by NASA as being able to enhance safety during four phases of flight: ground operations between gate and runway, departure operations, en route operations, and arrival and approach operations (Williams 2001). One study completed in 2003 by Arthur, Prinzel, Kramer, Bailey, and Parrish showed that 30 percent of all fatal accidents worldwide are categorized as CFIT, "where a mechanically sound and normally functioning airplane is inadvertently flown into the ground, water, or an obstacle, principally due to the lack of outside visual reference and situational awareness" (2003 146). This study focused primarily on adapting SVS technology to current raster PFDs installed in most modern commercial aircraft. Upon making the necessary modifications in aircraft simulators, it was shown that pilots flying the modified aircraft experienced significantly higher situational awareness and lower workload versus pilots flying non-modified avionics systems (Arthur et al. 2003). This study also showed that the integration of terrain data on smaller PFDs within older glass cockpit environments still increased SA and reduced mental workload significantly, although the larger displays on newer aircraft, such as the Airbus A380, showed a greater improvement (Arthur et al. 2003). The ease in which a SVS was integrated into current technology, combined with the benefits shown through the results of this study, make it difficult for Boeing and Airbus to justify not conducting more research into adding SVSs into their current product line. In addition, due to the product offerings already in the market from Honeywell and Rockwell Collins, SVSs could start providing benefits to commercial fixed-wing operations almost immediately, with very little additional training required for use of the system by current commercial pilots.

Using SVS and EVS in Conjunction with One Another

Another focus of research has been integrating both a SVS and a Forward Looking Infrared Radar (FLIR) EVS with the goal of further increasing situational awareness. In 2005, tests were conducted on a retrofitted Gulfstream G-V aircraft to see what combination of SVSs and FLIR technology integrated into HUDs and PFDs provided the largest increase in situational awareness and lowest mental workload (Arthur, Kramer, and Bailey 2005). It was proven that the integration of a SVS into a HUD was the best option for increasing situational awareness while adding FLIR data to a HUD was shown to do little more than increase clutter and pilot distraction. This has been attributed partially to the clean look of SVS data on a HUD, since everything above the outline of the terrain is left blank. A FLIR, on the other hand, displays everything it picks up within the IR spectrum on the HUD (Arthur et al. 2005). This further proves that the use of terrain databases and SVSs helps to decrease the chance of CFIT events by continually increasing pilot situational awareness.

Although not discussed in the previous study, a separate PFD with FLIR data may be greatly beneficial in order to augment the information pilots receive from the onboard terrain database integrated into the SVS. The primary method of collecting data for terrain databases to use topographic mapping from space, such as the 1994 Shuttle RADAR Topography Mission. This mission provided terrain data for 80 percent of Earth to a resolution of fifty-three feet (NASA 2000). This data is augmented with LIDAR topography data where possible, having a resolution of approximately one foot. This data is still very limited due to the intensive nature of data collection. Even with one foot resolution, LIDAR would not be able to detect various items, such as power lines and telephone poles, both of which are hazardous to the safety of an aircraft. By having FLIR data augmenting a SVS, a pilot could more easily see other obstacles, such as wildlife or other aircraft on the same approach path. Although the data is hard to interpret at times due to clutter, the incremental benefit of having the data accessible to pilots when needed will most likely be greater than its incremental cost of additional workload.

Pathway Guidance in the Prevention of CFIT

Pathway guidance, when used with SVSs, has been shown to help further reduce CFIT events. A study by Kramer, Prinzel, Arthur, and Bailey in 2004 showed that by having fly-through gates superimposed on a HUD with SVS terrain data, mental workload was further reduced and situational awareness increased, although there was no increase in flight path accuracy (2004). Even though there was no effect on flight path performance, this technology still offers great promise in increasing aviation safety over the next ten years. By making it easier for pilots to perform the regular duties of flight, it will allow the pilot to react more quickly to other situations. For example, if an aircraft had a landing gear failure on approach, a pilot would be able to focus more on troubleshooting the problem without having to worry as much about the flight path or altitude. This system could be improved further by integrating software into the pathway guidance system that would alert pilots with an audible and visual alarm if they were deviating from the intended path. It would be important to make the alarm easy to override due to the pathway projected on the HUD being dependent on what the pilot enters into the autopilot system. If the pilot is instructed to deviate from the intended path by ATC and did not have enough time to enter the data into the autopilot, the pathway guidance system would not have this information and would create a false positive. If it was difficult to override the alarm, this could create undue stress in the cockpit and could lead to pilot errors.

Adaptations

Adaptation of Synthetic Vision Technology for Ground Operations

Although air operations is where most fatal accidents occur, the ultimate goal of increasing safety of aviation can still be achieved by focusing also on ground operations. Current research is attempting to find ways to adapt SVS technologies to help reduce runway incursions by using a PFD with a top-down view alongside another PFD with a synthetic view of the airport. In 2006, Vernaleken, Mihalic, Guttler, and Klingauf conducted a study of a Surface Movement Awareness and Alerting System (SMAAS), which is comprised of the components listed above and a live data link with air traffic controllers (2006). This data link provides all data regarding the aircraft's assigned taxi route along with confirmation of whether the pilot is cleared to taxi across runways. Of the pilot's tested with this system, 100 percent agreed that it increases situational awareness on the ground for pilots, and 93 percent state that this system is a definite improvement over other systems they have today (Vernaleken et al. 2006).

As the SMAAS is further developed, it will allow for more efficient taxiing, fewer runway incursions, and the possibility of taxiing in lower visibility than what is currently possible. One downfall to this system is the inability to see other traffic taxiing on the surface, as the SVS shows only the permanent features of the airport. One solution to this would be to augment the PFD with data from a LIDAR or FLIR EVS. This would allow for other aircraft and vehicles to be seen in zero-visibility conditions. One study has already begun to look into the possibility of placing this EVS data on a HUD or head-mounted display, in conjunction with SVS data. It has proven to be highly effective in reducing ATC verbal error when used along with pilot interpretation of the verbal directions, while avoiding other adverse taxi events, such as aircraft coming nose to nose on a taxiway (Arthur, Prinzel, Shelton, Kramer, Willaims, and Bailey 2007). The main issue that has arisen is that the HUD seems to get cluttered with information and pilots feel that their workload is increasing. This is further compounded by the current lack of customization available for HUDs (Arthur et al. 2007). By implementing a simple clearing function, or allowing pilots to choose what data they want displayed on their HUDs, this technology could become invaluable. A clearing function is much easier to develop than the design of an entirely new technology that can bring the same benefits as a SVS-EVS hybrid system.

Advancements in Synthetic Vision Systems for Helicopters

Although helicopter flying is usually conducted at a much lower altitudes than other fixed-wing operations, the same concepts of SVS and EVS technology can be adapted to meet its needs. U.S. helicopter accident statistics, collected by Helicopter Association International, show that 6 percent of all helicopter accidents between 1997 and 2006 were caused by wire strikes (Helicopter Association International 2007). In fact, the data shows that over this same period, the total number of accidents in IFR conditions was less than that of wire strikes, meaning that at least 33 percent of the time, a lack of situational awareness led to a pilot flying into a wire during VFR conditions (Helicopter Association International 2007).

Due to the low flying nature of helicopters, it has been shown that a strong emphasis must be placed on creating high fidelity terrain databases based on LIDAR collection methods. While fixed-wing aircraft tend to fly low only on the approach and departure phases of flight, a helicopter conducting a pipeline inspection might never climb above one thousand feet during a three-hour mission. The traditional data used from the Shuttle RADAR Topography Mission, discussed earlier in this paper, does not provide enough precision for helicopter flight due to it being accurate only to within fifty-three feet (NASA 2000). This is especially true in mountainous areas. By constructing terrain databases with low resolution data, helicopter pilots may experience a CFIT event in low-visibility conditions, even though they believed they were fifty feet above ground level. One study conducted in 2006 by Sindlinger, Meuter, Barraci, Guttler, Klingauf, Schiefele, and Howland showed that using LIDAR terrain data of the mountains around Zurich provided a highly accurate synthetic view for helicopter pilots to fly with (2006). This study also discovered that an additional challenge for helicopters is the loss of a navigational fix. Since helicopters can fly in between mountains, it was difficult at times to maintain a constant GPS fix (Sindlinger et al. 2006). A constant navigational fix is required because, with no live location information, the SVS will not know what terrain to display. If this problem can be addressed by some other means, such as integrating an inertial navigation system for use when a GPS signal is not available, SVSs promise to be a highly useful tool in improving helicopter flight safety in times of low-level flight.

Although LIDAR terrain data will help prevent accidents involving helicopters impacting the ground, it still does not address the issue of wire strikes, due to its maximum resolution being approximately one foot. Yonemoto, Yamamoto, Yamada, Yasui, Tanaka, Migliaccio, Dauvignac, and Pichot are currently working on a system that integrates a color camera, FLIR, and millimetre-wave RADAR to help detect long and thin obstacles, such as power and other transmission lines (2006). Their goal is to provide an all-weather collision warning system for helicopter pilots flying at low levels. While these instruments and supporting software have been shown to see power lines up to 800 metres away, the imagery provided to the pilot is very rudimentary and hard to interpret, especially when considering that the pilot will be flying the aircraft at a low level while viewing the data provided.

Currently, the U.S. Army is conducting research in integrating 3D terrain elevation sensor data with onboard terrain databases that currently comprise SVSs. The goal of this research is to ensure database integrity while constantly increasing safety of low-level flight (Szoboszlay, Jennings, and Tiana 2006). Although this research is similar to the previously discussed obstacle avoidance sensors, it has the same downfalls as the previous study, in the sense that by showing the raw imagery from FLIR and millimetre-wave RADAR, the PFDs and HUDs become quite cluttered, leading to higher pilot mental workload (Szoboszlay et al. 2006). A possible solution to these problems would be to have a power-line collision warning system running in the background at all times with audible and visual warnings given to a pilot when in a situation requiring the alteration of the helicopter's present flight path. This would be quite similar to a TCAS system in the sense that it provides only conflict resolution commands, such as "climb," but not for times where everything is normal, when the command would simply be "continue on course." By the pilot needing to focus only on the SVS, it will increase SA and decrease workload, while providing the additional obstacle avoidance that the previously discussed systems specialize in.

Future Adaptations of Synthetic Vision in Pursuit of Aviation Safety

As history has shown before, with the co-development of the hydrofoil and airfoil in the late 1800s, there are often many spin-off effects in regards to the development of a type of technology, especially when two are simultaneously co-developed. With this in mind, it is assuring to see that SVS and EVS technologies are currently being adapted for the use in an ATC environment, along with the new Lunar Lander being developed by NASA.

Research is being conducted on the feasibility of integrating an EVS into an ATC tower environment. The primary source of information for ATC during ground movements of aircraft are visual cues they receive looking out the window. During times of fog, rain, and low visibility, these cues are all but missing, leaving the controllers directing aircraft and other ground vehicles blind, unless a ground RADAR system is installed. Ruffner, Deaver, and Henry have been researching the feasibility of adapting infrared (IR) sensors to restore visual cues to tower and ground controllers (2003). The research study identified many hurdles that must be overcome, such as developing a display medium for controllers that does not interfere with other duties, mounting the IR sensors, along with receiving instantaneous sensor data, considering the controller's current field of view and field of regard (Ruffner, Deaver, and Henry 2003).

Another option may be to combine this information, along with synthetic vision data, onto one display that will allow controllers to not only see the aircraft and other vehicles on the airport surfaces, but also the various buildings and other landmarks that controllers have become used to seeing during day VFR conditions. By developing a system that would allow all of this information to be presented on a form of head-mounted display, aviation safety would continue to grow, due to safe and efficient ground movements provided by air traffic controllers. In addition, this technology could further increase safety inside the cockpit by being adapted to provide an unlimited and instantaneous field of view to pilots during periods of low visibility.

Although not directly related to aviation safety, the challenges faced by NASA's implementation of a SVS for lunar and planetary landing vehicles are similar to those faced by SVSs in the cockpit today. The main technical challenge to implementing a SVS for lunar landing vehicles is creating an accurate enough database that will provide lunar vehicle pilots enough fidelity to facilitate safe approaches to the moon's surface (Williams, Arthur, Shelton, Prinzel, and Norman 2008). As this technology is further developed, there is a possibility that it may be applied to creation of a terrain database for Earth. Since aircraft over-flights with LIDAR data collection systems are not possible over the surface of the moon, any methods developed to help with the creation of an accurate topography database for the moon may be applied to Earth data collection. In turn, this may present a lower cost and more feasible approach to collection of this vital data for terrestrial SVSs.

Safety Issues

Synthetic Vision Makes Aviation Safer

Although SVSs work to increase SA, decrease workload, and have the potential to increase safety of aviation, there is still work to be done. For SVSs to be effective, the terrain database on which it is based must be of high quality and complete, especially for helicopter low-altitude operations over rugged terrain. If pilots are to rely on a technology, it must be thought to be effective and accurate 100 percent of the time.

Arguments made by Boeing and Airbus stating that there is no place for synthetic vision systems in the modern commercial aircraft cockpit are baseless. Although current navigation systems do allow for safe travel through air corridors, only 9 percent of fatal air accidents on fixed-wing aircraft occur during the cruise phase of flight (Boeing 2008). By implementing SVSs on commercial aircraft now, both Airbus and Boeing would be taking a proactive approach in the attempt to prevent fatal air accidents from occurring in the future due to CFIT. It was only after accidents, such as these, that the current GPWS system was introduced. It is not a matter of *if* another CFIT event will happen; it is a matter of *when*. No other technology currently in development offers a larger possibility of increasing safety within the realm of aviation during the next ten years.

Aviation Technology and Safety Soars to Lofty New Heights on the Wings of the Super Jumbo

By Rupert Schuld

Aviation has seen many developments in its relatively short century-old history, including the concept of the Super Jumbo or Ultra High Capacity Aircraft (UHCA). The idea of a double-deck aircraft with multiple engines, carrying over 600 passengers, was fostered during the recession of the 1970s when a need developed for larger, more efficient aircraft. Many super jumbo concepts from various aircraft manufacturers were developed; however, the massive cost of development and inadequate technology prevented these aircraft from passing out of the design phase. It was not until the twenty-first century that Airbus would unveil the first Super Jumbo, the A380. This aircraft is far ahead of its time and will prove to transform the aviation industry. Many new systems addressing safety have been developed and improved, including:

- Multi-role/slide life rafts

- Landing gear and the new Brake-to-Vacate braking unit

- High tech materials and processes such as glass fiber reinforced aluminum laminate (GLARE) and laser beam welding (LBW) that will change the way aircraft are constructed

- Cockpit technologies including the control cursor device (CCD) and aircraft environment surveillance system (AESS)

Furthermore, the A380 is addressing the issues of global warming, aircraft congestion, and air traffic control constraints. Finally, the large size of the A380 is posing issues for major airports to accommodate this aircraft. Super Jumbos will create a safer, more efficient aviation industry as a result of innovative technologies and manufacturing processes.

History of the Super Jumbo

The Middle East oil crisis in the 1970s sparked a global recession which crippled many airliners and forced aircraft manufactures to design more efficient aircraft that would address the problem of sky-rocketing fuel prices. In the early 1980s, the appeal to airliners and aircraft manufacturers for larger wide-body aircraft and the concept of the super jumbo became stronger. The most important design element of an aircraft is its cross-section and the ability to hold increased passenger loads, as well as standard LD3 cargo containers (a pallet used to store large amounts of payload on wide-body aircraft into a single unit), reducing ground crew time and overall efficiency (Norris and Wagner 2005). Boeing was the first to address this problem by implementing the design and construction of the 747, with McDonnell Douglas and Lockheed closely following with their variants. Airbus entered the scene trailing with the implementation of the twin aisle (TA) TA 9 and TA 11 projects, which later became the A330 and A340 options (Norris and Wagner 2005). With all major passenger aircraft manufacturers established in the market with large capacity aircraft, fierce competition was being experienced on a higher level.

Boeing came out with improved versions of the 747, while Airbus countered with improvements in the A340. The advantage Airbus had over American aircraft manufacturers was adaptability and innovation, which was largely driven by the need to do more with less. The A330 was a twin engine aircraft and the A340 was a four engine aircraft like the 747, but both aircraft utilized a common wing and fuselage cross-section allowing the company to be more profitable (Norris and Wagner 2005). The resulting flexibility of the A340 series resulted in Airbus's chief engineer, Jean Roeder, announcing the need for a new Ultra High Capacity Aircraft (UHCA) or Super Jumbo to satisfy the need for links to dominant hubs. The major constraint of most manufacturers to create larger, more efficient aircraft was cost; however, Airbus was going to create an aircraft using existing products, like the A340 and A330, and their advanced technology (Norris and Wagner 2005).

The first super jumbo concept in 1988 was similar to the A340, but dubbed Horizontal Double Bubble (HDB), which basically stretched the existing A340 laterally, melding two of its fuselages side by side. It also used existing A340 components, including wings, tail, and cockpit, resulting in an aircraft 20 percent larger than Boeing's jumbo jet itself. This aircraft would operate at costs 15 percent lower than the 747-400 (Norris and Wagner 2005). However, the ingenious idea of using existing aircraft parts of the A340 faltered, with poor results in the wind tunnel testing phase. Boeing later entered the Super Jumbo arena in 1991 after United Airlines requested a 650-seat trans-Pacific aircraft called the N650. Boeing forecasted at the time that 54 percent of the value of the commercial market up to 2005 would be for 350+ seat aircraft, with the number increasing to 500 seats in 2020 (Norris and Wagner 2005). They believed the 747 and 777, along with a Super Jumbo, would meet this demand. Airbus responded with their A2000 concept, and now United Airlines was faced with two possible Super Jumbo options. Both Boeing and Airbus wanted to mitigate their losses and conducted a study of the major world airlines, assessing their need for a Super Jumbo aircraft. Many expressed interest, and many rejected the idea, including Qantas Airlines of Australia. Boeing, unlike

The "8" in A380 was chosen because it visually resembles a cross-section of the aircraft's double decks; "8" was also picked because it represents good luck in Asian countries.

Airbus, was faced with the challenge of taking advantage of the well-established 747 family and making larger variants, dubbed the 747 X Generation (Norris and Wagner 2005).

Airbus continued with its super jumbo concept. They realized the need for these aircraft in regions not limited to, but including China, where air traffic is growing by 20 percent per year internally and 12 percent internationally (Norris and Wagner 2005). The majority of China's aviation industry is operating old Russian aircraft that are due for retirement. The influx of Super Jumbos made by established Western manufacturers would only improve aviation safety in a country crippled with relatively high accident rates. Improved efficiency will also be experienced and, being one of the largest economies and trade partners in the world, will be beneficial on a global scale. The Super Jumbo will allow more cargo to be carried at a single time, reducing the lead time traded goods currently experience. Much to the world's surprise, McDonnell Douglas was the first to emerge with a firm new double-decker design called the MD-12X. However, McDonnell Douglas was the victim of bad timing and unveiled its super jumbo concept in 1992, in the midst of one of the worst recessions the airlines had seen (Norris and Wagner 2005). Orders were too few or cancelled, and McDonnell Douglas soon faltered. All these attempts at constructing the Super Jumbo failed, mainly due to high investment costs and the lack of availability of adequate technology. However, advancements in technology saw aircraft manufacturers construct larger, safer, faster, and more efficient aircraft. Airbus eventually became the only manufacturer to pursue the Super Jumbo and achieved success with its latest concept, the A3XX (Norris and Wagner 2005).

The A3XX first flew on 27 April 2005 and was dubbed the A380, as the "8" represented the double-deck fuselage that clearly identified its Super Jumbo status. This double-deck concept was a defining feature from the start when Super Jumbos first appeared on the drawing boards, and ever since has been a determining factor in whether an aircraft is a Super Jumbo. The number was also picked because it represented good luck in Asian countries for which the A380 was targeted (Norris and Wagner 2005). The A380 could seat over 800 passengers in a high-density single class, and 555 passengers in a three-class layout (Airbus 2009). The massive Super Jumbo specifications include a length of 239 feet and a wing span of 261 feet (66 feet longer then the 747), and four powerful engines (Airbus 2009). It was the first Super Jumbo of its kind to enter the commercial aviation industry and proved to the world that it combined the very latest in materials, systems, and manufacturing processes which have and will continue to transform the future of aviation safety and efficiency.

Safety Systems
Multirole Slide/Life Rafts

New comprehensive safety systems, such as the multirole slide/life rafts, were design drivers from the start of the A380 program and will greatly influence aviation safety and efficiency in the future. For speedy evacuation in an emergency, Airbus treated the A380's double decks as two totally independent passenger decks and restricted the velocity of passengers at the bottom of the evacuation shoots to no greater than those certified on a single-deck aircraft (Norris and Wagner 2005). The most challenging aspect of the A380 was the main deck 3R and 3L exits that were located over the wing. These exits still required an inflatable device not common to modern-day aircraft that would allow passengers to escape back over the trailing edge of the wing behind the upper escape slides towards the rear of the aircraft. Depending on the configuration of the aircraft during an accident, the inflatable slides have sensors that detect if the aircraft is high or low and adjust the length of the slides by thirteen feet to facilitate the most efficient evacuation (Norris and Wagner 2005). This will drastically reduce the number of injuries passengers will experience when evacuating an aircraft after an emergency.

Many injuries in past accidents are a result of evacuation slides not reaching the ground because of the aircraft's position after the incident or accident. Examples include the incident at Gimli, Manitoba, in 1983 when

Air Canada flight 143, a 767, ran out of fuel over Manitoba and successfully attempted a forced approach on a drag strip with the nose gear retracted (TSB 2006). This resulted in the aft emergency slides not reaching the ground and thus injuring many passengers during evacuation. Features like this on A380 Super Jumbos are transforming the way evacuation slides are installed in modern airliners and will increase safety and efficiency in emergency procedures well into the future.

Landing Gear and Braking System

Another feature of the A380 that will greatly influence the future of aviation safety and efficiency is the landing gear and braking system. The landing gear of the A380 is unconventional in terms of modern-day airliners and incorporates titanium construction and a double-puck braking system yet to be seen in the aviation industry. These advancements alone require only the use of inboard thrust reversers, and together the A380 has a stopping distance similar to the much smaller A340 and better than the 747. The elimination of the outboard thrust reversers actually improves efficiency and safety. A total of 7700 pounds is saved in aircraft weight, and the elimination of complex thrust reverser components means that future Super Jumbos will be less likely to have something go wrong. Such was the case of Lauda Air Flight 004, a 767, whose thrust reversers deployed during cruise flight, resulting in a subsequent crash, killing all aboard (Sogame 1993).

To improve safety and efficiency further, the A380 is the first aircraft to use the Brake-to-Vacate (BTV) system, which allows pilots to use the auto-flight, flight controls, and auto-brake systems to regulate deceleration after touchdown. The BTV system enables the A380 to reach a specified exit ramp at the correct speed under any conditions increasing the efficiency of airports to handle aircraft as a result of growing congestion conditions (Airbus 2009). This landing gear and braking advancement will improve performance of future large aircraft and quite possibly lead to a reduction in landing overrun accidents that comprise the majority of aircraft incidents around the globe. By eliminating heavier and mechanically advanced aircraft components, such as outboard thrust reversers, coupled with newer advanced systems such as the BTV, the A380 will provide the necessary technology to increase aviation safety and efficiency for years to come.

New Materials and Manufacturing Processes

As a result of the incorporation of high-tech materials and processes in the construction of the A380 Super Jumbo, many advancements in aircraft efficiency and safety have been achieved, and this will transform the way aircraft are constructed in the future. The A380 is the first aircraft to be constructed with more than 40 percent of its components made of carbon composites and advanced metal alloys (Proctor 1994). Specifically, the A380 incorporates a one-piece wing structure with a carbon fibre wing box, which is nearly 30 percent stronger than any other commercial aircraft (Norris and Wagner 2005).

Airbus also uses glass fibre reinforced aluminum laminate or GLARE on its Super Jumbos; this is much stronger than conventional aircraft aluminum and reduces aircraft weight between 15 and 30 percent (Cefrey 2002). This material also has better fatigue, corrosion, and damage-resistance properties. In an airline economy where margins are small, GLARE will allow airliners to save on fuel costs and gain larger profit margins. This will also provide a safer, more efficient aircraft, as maintenance overhaul time will be reduced as a result of increased product quality and construction.

The aviation industry may experience a complete restructuring of aircraft maintenance check times and procedures as aircraft manufacturers will certainly adopt the materials used in the construction of the A380 Super Jumbo. To conjoin the various components of the Super Jumbo, the new advanced manufacturing process of Laser Beam Welding (LBW) is used. This method mainly attaches the stringers of the lower fuselage shell section, however, it is used throughout the whole construction process (McCormick 2003). LBW saves weight by reducing the weld material and includes an independent inspection unit as the parts are welded, leaving no faults whatsoever. This will greatly improve the safety and efficiency of the aircraft construction process, as the inspection unit will be able to detect faults that human beings are not capable of detecting. Furthermore, LBW will provide a more structurally reliable aircraft, as these welds fuse aircraft components more accurately and thoroughly then the current human welder, and will no doubt be adopted by aircraft companies in the construction of future aircraft. This process eliminates the need for fasteners found in modern-day airliners, which are a major source of corrosion and fatigue.

The A380 is designed to operate for 140,000 flight hours, 19000 cycles, or a service life of twenty-five years. It is the most ruggedly structured aircraft operating in the world. All this new technology will inevitably make future aircraft stronger, lighter, and more reliable, enhancing their overall safety and efficiency.

Cockpit Automation

One of the most distinct features of the A380 that will impact the efficiency and safety of aviation in the future is the highly automated cockpit. The flight deck of the massive A380 super jumbo has similar features to existing airbus aircraft. The purpose of this is to make the transition easier and quicker for new pilots already trained on existing aircraft. This will increase efficiency and reduce the down time airlines have to spend on training pilots (Sutton 1997). Airbus studies reveal that experienced pilots become accustomed to the cockpits of the aircraft they operate. Over time flight crews begin to interact with the cockpit instinctively and subconsciously know where all the features of the flight deck are located. By preserving the original airbus layout, pilots will transition smoothly and operate the aircraft with a statistically lower rate of being involved in an accident or incident (Airbus 2009). However, there are many features of the flight deck that have never been seen before and that will transform the way cockpits are designed in the future. One of them is the cursor control devices (CCD), which are peculiar-looking bulges on the pedestal beside the pilot. The CCD is designed to operate like a computer mouse in the roughest turbulence and make the flight displays more interactive for the pilot (Norris and Wagner 2005). They can even steer the aircraft using the CCD by clicking on a direct-to command that interrelates with the flight management computer to direct the aircraft to a new way point. This will make navigating more efficient, allowing the crew to focus on more critical systems and enhance productivity and ultimately safety throughout the flight.

New display features of the A380 Super Jumbo will also help enhance the safety and efficiency of aviation in the future. A total of eight 6-inch-by-8-inch large-format liquid crystal displays (LCD), including two primary flight displays (PFD), two navigation displays (ND), two multifunction displays (MFD), an engine/warning display, and a single systems display, are present within the flight deck (Norris and Wagner 2005). The A380 also holds the most advanced LCD digital heads up display that will be used during precision approaches in instrument meteorological conditions (IMC). A Honeywell aircraft environment surveillance system (AESS) acting together with the eight displays can distinctly show the aircrafts altitude profile relative to the ground (Norris and Wagner 2005). The projected flight path will be displayed in profile and will immediately warn the flight crew aurally and graphically if the planned flight path will put the aircraft in harm's way. The AESS will also incorporate traffic alert and collision avoidance (TCAS), enhanced ground proximity warning (EGPW), and weather radar to help refine the overall surveillance picture. For example, the AESS uses EGPW data to adjust the angle of the weather radar sensor to eliminate terrain clutter along the intended flight path and optimize the scan (Norris and Wagner 2005). This will allow the crew to gain a 3D perspective of the environment they are operating in, including all the hazards that may be present in the Super Jumbo's surrounding airspace. This will improve situational awareness of the crew and allow them to make more informed decisions regarding the safety of the aircraft. A reduction in aircraft accidents, especially during the approach phase, where many accidents have been a result of controlled flight into terrain, should be significantly reduced. The automated cockpit of the A380 Super Jumbo will inevitably be incorporated into future commercial aircraft, increasing the safety and efficiency of the aviation industry.

Environmental Impacts

The A380 Super Jumbo uses four of the most powerful and greenest engines, and will impact the future of aviation greatly as the world addresses the issues of global warming on a greater scale. Each American-built GP7200 engine is rated at 70,000 pounds of thrust and the A380 has the option of providing a total combined thrust of 336,000 pounds, which makes it the most powerful airliner in the industry (Singleton 2008). However, the world is addressing the effects of greenhouse emissions more critically, and many have expressed deep concern over the environmental impact an aircraft the size of the A380 can have. The A380 Super Jumbo has a total range of up to 9,320 miles and offers better fuel economy per passenger mile than most small hybrid passenger cars, burning only 3.1 litres of fuel per seat every 100 kilometres the aircraft travels, equating to 20 percent less fuel consumption than the next largest airliners (Singleton 2008). Furthermore, the Super Jumbo burns 12 percent less fuel per seat then its closest competitor, the Boeing 747, while at the same time carrying 35 percent more passengers, increasing efficiency, reducing operating costs, and minimizing the effects Super Jumbos have on the environment (Airbus 2009). The impact that these engines will have on the aviation industry will be significant. Many future aircraft will incorporate this engine's technology (albeit on a different power scale), to address the growing pollution issues and provide a greener option in a world that is looking for better alternatives. The Super Jumbo proves that air travel can be much more efficient and will be the next best substitute in transportation, greatly impacting the future direction of the aviation industry.

Impact on Emerging Markets

The A380 and future Jumbos are going to fill a demand, particularly in Asia, as a result of urbanization and global economic development. It is estimated that by 2020 there will be sixteen cities in the world that are going to have populations greater than twenty million; ten of these cities will be in the Asia-Pacific region (Lawrence 2006). It is estimated that in China alone, twenty million new middle-class consumers are rising annually in new intercity developments along the east coast. This is increasing the amount of available disposable income individuals are willing to spend on local and international air travel (Lawrence 2006). The fastest growing international air routes include Pan-Asia, Asia–U.S., and Asia–Europe. This is creating demand for routes to major population centres such as New York, London, Frankfurt, Rome, Los Angeles, Amsterdam, San Francisco, and other large cities that are in close proximity to major hubs (Lawrence 2006). These consumers alone will dramatically increase the concentration and route density of the commercial aviation industry. Statistical evidence indicates that Asia will have to increase the number of large capacity jets by 368 percent by 2020 (Norris and Wagner 2005).

The A380 Super Jumbo is the perfect aircraft for this dilemma because of its size, operating economics, and range. It is estimated that by the year 2015, three Chinese cities; Beijing, Shanghai and Guangzhou, will surpass a population of forty million. These three cities will account for 90 percent of U.S.–China travelers' destinations of choice (Lawrence 2006). There are also numerous routes that have been growing rapidly in density during the last decade, including London–Chicago, London–Dubai, and Paris–Rome. Currently, these routes are almost at their saturation point in terms of congestion. However, the A380 and future Super Jumbos will be able to increase passenger density on these routes and alleviate the number of flight movements. This will help increase the efficiency and safety of the commercial aviation industry in the future by reducing the workload demands on ATC.

Impact on Air Traffic Congestion

Air traffic congestion severely constrains the efficiency, safety, and capability of the Air Traffic Control network, and it is having an increasingly difficult time in handling aircraft approaches into major airports and facilitating gates on the ground. Since the cancellation of the N650 Super Jumbo program by Boeing, it is believed that medium-sized jets on point-to-point networks are going to be the wave of the future (Norris and Wagner 2005). However, with proven statistical evidence, it is hard to imagine that the Air Traffic System could handle an additional 6200 medium-sized twin aisle jets (as are predicted to be needed by Boeing), as problems are going to arise in terms of capacity constraints and environmental and security issues (Norris and Wagner 2005). The new concept of free flight hopes to loosen the control of aircraft routings, however, as a result of the 9/11 terrorist attacks, it is hard to believe that countries will allow a reduction of authority over flight routings. Therefore, the Airbus A380 and future Super Jumbos will inevitably be the solution to this growing problem, resulting in the greatest impact on aviation in the future.

Airport Compatibility

As the A380 and future Super Jumbos aim to solve the problems of congestion, they will be useless if airports cannot accommodate them. The impact that the Super Jumbos will have on airports in the future will be significant. After initial reviews by Airbus of seventy runways worldwide at key destinations, seventeen would require modification of taxiways an additional twenty-three feet and runways to a width of 200 feet (Norris and Wagner 2005).

Airbus convinced sceptical major airport operators that it would cost less if they made more efficient use of runways and gates to adapt to UHCA, rather then constructing more runways and gates to handle larger amounts of smaller aircraft (Airbus 2009). Airbus estimated the cost to be small in comparison to the $20 billion major airports already spend on upgrades. The U.S. alone would have to spend only $520 million to prepare for future Super Jumbos (Fife and McNerney 1998). The majority of the upgrade costs at airports would be to extend taxiways and runways. However, the A380 has a relatively equal maneuvering area in comparison to the 747, which can be attributed to its rear-rotating six-wheel body landing gear (Spencer 2005). Moreover, the A380 is fitted with closed-circuit TV cameras mounted in the empennage and belly to help prevent the crew from taxiing over the edges of runways and taxiways. As a result, airports would have to extend the shoulders of taxiways and runways only and not the complete maneuvering area (Spencer 2005).

Weight was another concern for airports, but after several static tests, Airbus confirmed that the Super Jumbo would pose no issue, as its Aircraft Classification Number (ICAO method of calculating load factor) would be from 66 to 72, significantly less then the 83 of Boeing's 777 (Norris and Wagner 2005). In light of this information, major airports have decided to go ahead and prepare for the future Super Jumbos.

The Super Jumbo is transforming the way airports are built. For example, London Heathrow is completely rebuilding Terminal 3's pier 6 to provide four double-deck A380 stands at a cost of $180 million and a further seven at terminals 4 and 5 by 2011 (Norris and Wagner 2005). Also, the Doha International Airport in Qatar will be designed and built specifically for the double decker. The airport will be expanded an additional 40 percent, with final completion in 2015 at a cost of $5.5 billion, allowing for the movement of 50 million passengers annually (Norris and Wagner 2005). Airbus has proven the capability and operation of Super Jumbos at major airports to be feasible, and, as the number of Super Jumbo's entering service increases, so will the development of airports that cater to them.

Conclusion

The A380 and Super Jumbo successors will have a profound effect on the future of aviation and will provide the technology for a safer, more efficient industry for years to come. The advanced safety systems that include the multirole slide/life rafts, landing gear, and BTV will drastically improve the future capabilities of aircraft and should reduce the amount of injuries, incidents, and accidents that current aircraft are involved in. The advancement in high tech materials and processes as a result of the A380 will allow future aircraft to become more efficient and safer, transforming the way maintenance will be conducted on aircraft in the future. The highly automated cockpit of the A380 will provide a better perspective of the external environment for pilots, improving situational awareness, as well as allowing the crew to focus on more critical systems, thereby enhancing the safety and efficiency of future aircraft that will implement this technology. The most powerful combined thrust available to date, of the A380 engines will prove to be the most economical as well. This advanced engine technology will allow the aviation industry to continue to grow, without contributing as much harmful emissions as other modes of transportation. The A380 will improve the aviation industry by solving the growing problem of world aircraft congestion, alleviating the constraints of air traffic control. The growing concern of airports to accommodate the A380 is relatively insignificant due to the advancements made in aircraft technology as a result of its development. It will surely transform the way airports are improved, constructed, and designed. Overall, the A380 and future Super Jumbos will have the greatest impact on aviation in the next ten years and will inevitably transform various aspects of the commercial aviation industry in terms of efficiency and safety.

Blended Wing Body (BWB) Aircraft
By Ryan Clarke-Boles

Over one hundred years of aviation has been explored since 1903, when the Wright Brothers first took flight. However, in this time, the world has only scratched the surface of humanity's quest to conquer the skies. The recent buzz in the aviation world, as of 2009, has been on the production of the Blended Wing Body by Boeing, NASA, and major U.S. engineering universities.

The Blended Wing Body is a hybrid aircraft which combines the features of traditional airline aircrafts and features of the flying wing designs. This unique design enables almost all of the airplane's body to produce lift while also reducing drag. The Blended Wing Body will be capable of seating upwards of 1000 passengers, 450 more than the current Airbus A380 holds, and will operate at speeds and distances similar to that of traditional airline aircrafts.

The BWB design has remarkable specifications that, along with its unique shape, will help to cut down massively on operating costs and improve other competitive aspects of airlines.

Advantages
Fuel Efficiency
Fuel is the highest operating cost of airline operation, making up approximately 30 percent of all operating costs (Air Transport Association 2009 Chart 1). Studies have shown that the Blended Wing Body will be able to reach fuel burn 27 percent lower than its conventional Airbus 380 rival (Mizrahi 2009 ¶13). With jet fuel costing $1.63 per gallon, airlines are expected to save up to $1302 per hour flying the BWB (International Air Transport Association 2009 Chart 1).

The Blended Wing Body can boast such impressive fuel efficiency for a variety of reasons, the most important being its overall design. The integration of engines, wings, and the body into a single lifting surface to achieve maximum efficiency, which mean, less engine power and, hence, less fuel to reach the same amount of lift and speeds as conventional aircraft (National Aeronautics and Space Administration [NASA] 1997).

Noise Levels

This noise pollution factor is another key selling feature for the Blended Wing Body. Engineers at MIT and the University of Cambridge have helped design this aircraft to be virtually imperceptible outside the confines of an airport, thus allowing for public satisfaction and more route opportunities for airlines. They designed the airframe to be as smooth as possible, limiting noise caused by turbulent airflow along the surface (Massachusetts Institute of technology [MIT] and University of Cambridge [U of C] 2006). They have also incorporated noise dampening technology into the engines to greatly reduce engine noise, which is the largest contributor of aircraft noise.

Passengers

The Blended Wing Body will be able to enhance the passenger experience. First, with its unique windows, which are forward facing, passengers will see a similar view of what pilots see in the flight deck. Furthermore, it has an interesting new seating configuration, which separates the cabin into small rooms, allowing for a more private feeling.

(CONTINUED)

SAFETY IMPACT ON AVIATION INDUSTRY

The BWB project has focused just as much attention to safety as it has to any other aspect of this aircraft.

Pressurization

Critics believe that the structural integrity of the BWB will not be able to sufficiently withstand the immense force of pressurization at altitude. Most commercial aircraft cabins are internally pressurized up to 8000 feet, and the cylindrical design of the fuselage enables it to evenly distribute the pressure around the entire body. Researchers have solved this problem in the BWB by using ten intermediate front-to-back ribs to connect the upper and lower wing skins. These ribs separate the interior into ten passenger bays, which allow the pressure to be evenly distributed over the body (NASA 1997).

CO_2 Emissions

With aviation resulting in 20 percent more carbon dioxide emission than previously predicted, it is the onus of manufactures such as Boeing and Airbus to come up with a solution (Triple Pundit 2008). With 27 percent better fuel efficiency than the A380, the Blended Wing Body has a per-passenger fuel burn rate comparable to a fuel-efficient car, such as a hybrid. Furthermore, the Blended Wing Body's unconventional configuration may be the only way for airlines to meet aggressive future environmental demands from governments and governing aviation bodies like ICAO.

CONCLUSION

The Blended Wing Body will reach an unparalleled status, forever changing the face of aviation. With its radical new look and unique design, it has been intended to make air travel much more efficient. It will also continue to make air travel a safer means of transportation. Other than the overall design, the Blended Wing Body will also maximize the use of the newest technology, making it the most technologically advanced commercial aircraft ever built. Aviation has gone through a tremendous learning experience to create the Blended Wing Body, and this design is the next step in our ever-changing quest to conquer the skies.

New and Emerging Aviation Markets

By Colin Brown

Introduction

The aviation industry is rapidly changing and developing new ways of accommodating comfortable, fast, and efficient travel. A recent analysis of the global aviation industry suggests a growth rate per year of 5.6 percent until 2024 (Zinnov 2007 1). Notably, emerging markets such as China, the Middle East, and India contain the greatest opportunity for growth (1). All of this natural development is a part of any maturing industry, but aviation is unique. The aviation industry requires a tremendous amount of cooperation and coordination between nations who otherwise are very different and have radically different policies. As aviation markets continue to grow, these differences may create inefficiencies and safety hazards. A perfectly functioning aviation industry would have globally consistent standards and manage safety effectively. Four major inconsistencies between countries will be exaggerated as the aviation industry grows:

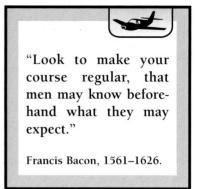

1. Language standards: English is the language of commercial aviation worldwide. However, significant growth is taking place in parts of the world where English is not common, such as in Asian markets. Therefore, the standard for English in aviation must be raised, with simple and concise standard phraseology.

2. Cultures: Differences in national culture have resulted in radically different corporate cultures among airlines across the globe. However, the most important culture should be standard for all aviators: the safety culture.

3. Training: There is significant room for improvement in aviation training across the globe. Therefore, abnormalities within aviation training should be standardized to meet a global requirement.

4. Regulatory powers: Currently, ICAO is the closest agency to representing a global governing body of aviation. However, the argument can be made that ICAO should be given actual regulatory powers to make decisions that benefit the entire industry.

Language Standards

Communication is becoming a frequent problem for pilots and air traffic controllers. Without a doubt, clear and concise communication will become increasingly important, as the world's airspace grows busier. English has been generally implemented as the language of commercial aviation. As Latin has been accepted as the language of medical terminology, English has been accepted as aviation's mother tongue. It makes logical sense for English to continue to remain the primary form of aeronautical communication since the system is built around the language. Also, economically, because of training and equipment modifications, it would be incredibly expensive to change the language. There is no question, English is here to stay unless some unforeseen development occurs. However, the quality and optimization of English remains under question. Safety is achieved through consistency; therefore, language is extremely important.

Consistent and simple phraseology is an easy way to eliminate some confusion from the skies. Simplification through replication would allow for pilots with English as a second language to cope with a lesser understanding of the language. Yet even in North America, where English is predominantly the language of choice, variations in phraseology exist. Until recently, Canada, the United States, and ICAO recommended different phraseology in their "Line Up/Line Up and Wait" orders (Change in Canadian ATC Taxi Phraseology 2008 ¶4). Fortunately, Canada decided to change their radio calls to become more consistent with ICAO (¶5). While ICAO recommended phraseology, "Taxi to Holding Position," meant taxi to but **not** onto the runway, Canada's air traffic controllers used "Taxi to Position" to order an aircraft to line up **on** the runway (Fox 2009 ¶5). As one can imagine, these similar statements could be easily confused during a busy time in the aircraft, such as takeoff. Simple miscommunications can have catastrophic effects in aviation and create inefficiencies (such as unnecessary go-arounds). This example demonstrates how inconsistent phraseology can lead to confusion for both English experts and learners alike. With 189 contracted States as members of ICAO, variances among common phraseology and terminology can easily make the consistent use of English very complicated (ICAO List of Contracting States 2006). The most proficient English-speaking pilots could find themselves in dangerous situations due to communication discrepancies, let alone a pilot relatively new to the English language.

The English discrepancies being experienced now are troublesome, but what does the future hold? ICAO has identified this issue as a serious concern, and has developed English proficiency requirements (ICAO Language Proficiency Requirements 2007). However, ICAO cannot force conforming States to comply. Unfortunately, the results of the proficiency requirements have not been impressive. For example, it is estimated by the Polish Civil Aviation Office that "only 15 out of 800 Polish pilots flying internationally have

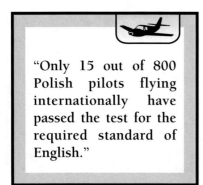

"Only 15 out of 800 Polish pilots flying internationally have passed the test for the required standard of English."

passed the test for the required standard of English" (Webster 2008 ¶2). For ICAO States struggling with the standard, there have been exemptions made available until March 2011 (Webster 2008). Recently, a Polish airliner was lost wandering the skies for nearly half an hour after coming "within seconds of colliding with another plane" (¶1). The Air Accident Investigation Branch discovered the main cause of this dangerous incident to be "the difficulty of obtaining information from the pilots because of their limited command of English" (¶8). This is an example of how poor English usage and comprehension can lead to hazardous situations. Although the issue with consistent and proficient English usage has been identified, it has not been adequately resolved. As countries continue to grow, and international carriage becomes more frequent, poor communication through the use of English will become more of a threat to aviation safety.

Culture

Culture varies from country to country. Different social norms and religious views can impact one's actions in daily life, including actions within the cockpit. The classic comparison is Asian relative to Western culture. It can be argued that Asian cultures are too obedient to authority. This can result in a silent co-pilot when the captain is clearly making an error. On the other hand, the Western mindset allows for more communication between crew members. To reveal how differences in culture affect safety, the example of Korean Air will be used in comparison to United Airlines from 1988 to 1998. During this time period, United Airlines lost a plane due to accident at a rate of .27 per million departures (Gladwell 2008). During that same time period, Korean Air's loss rate was 4.79 per million departures (180). Simple math reveals that flying on United Airlines during that time period was more than seventeen times safer! One might conclude that United Airlines, being an American company, spent more time flying in North America where aviation was more established. Furthermore, you could argue that navigation aids were more readily available, aircraft controllers and pilots were better trained, or equipment was more advanced.

Although some of these arguments may have relevance, the main contributory factor was culture. Nearing the end of this disastrous time period for Korean Air, Korean President, Kim Dae-jung, delivered a statement, "The issue of Korean Air is not a matter of an individual company but a matter of the whole country . . . our country's credibility is at stake" (182). Shortly after this speech, Korean Air made some incredible progress. Presently, Korean Air is a member of the SkyTeam alliance and has recorded a "spotless" safety record since 1999 (Gladwell 2008). The question is: what changed between 1998 and 1999 to make Korean Air suddenly more safety oriented? There was no significant change in staffing that would have resulted in the shift. Likewise, the equipment was mostly the same. Korean Air simply discovered the importance of changing its organizational culture by emphasizing safety above all else. The solution to Korean Air's dilemma was to support a safety culture. Korean Air's awareness shift towards a safety culture was enough to drastically improve its accident rate. No country's individual cultural practices can have precedent over safety in the global industry of aviation. Emerging markets, such as China, the Middle East, Brazil, and India, have very different approaches to authority and communication. However, the safety culture required for the aviation industry must take precedent over national culture while in the cockpit. Teamwork and communication remain the most valuable skills for completing a safe flight. There is no room in aviation for crew to disregard safety proceedings for fear of insulting one another through cultural unpleasantries. The global aviation industry must recognize the importance of "soft skills" (teamwork, communication, etc.) to prevent emerging markets from repeating Korean Air's mistakes.

Ignoring national borders, cultural problems exist within aircraft cockpits worldwide. In 1999, an aviation psychology study was conducted to determine what communication strategies were used among American Airlines' captains and first officers. Crew members were given the scenario that bad weather was ahead and they had to request that their partner avoid the weather (Fisher and Orasanu 1999). The requests made by the crew members were divided into six classifications: commands, crew obligation statements, crew suggestions, queries, preferences, and hints. Command was considered the most direct and orderly, while a hint was the most indirect and passive. The results showed that first officers predominantly used hints to notify captains of the oncoming bad weather (Fisher and Orasanu 1999). In contrast, the captains relied on commands to deliver their messages (Fisher and Orasanu 1999). Both situations resulted in poor crew resource management. The captain seemed to have a large amount of power, while the first officer remained underutilized. As growth occurs and airspace becomes more congested, the average air crew's task will be more complex and difficult in nature. Therefore, air crews of the future, regardless of nationality, or rank will be required to establish an effective and balanced safety culture on the flight deck.

Training

The foundation of aviation is effective training. A consistent training regime across the globe would eliminate many aviation abnormalities. Granted, this would require a tremendous amount of organization that has yet to be demonstrated on a global scale. The closest agency to providing consistent standards is ICAO. The ICAO Safety Oversight Audit Program found a "lack of standardization in state's requirements and practices" regarding licensing and training (ICAO's Review of International Requirements for Flight Crew Licensing and Training 2006 ¶13). This is an ironic finding since one of ICAO's main objectives is creating efficient and consistent standards. There is no excuse for this issue to exist. States are all attempting to create the same product—efficient and safe pilots. In most cases, the same aircraft and equipment are being used. Furthermore, the ICAO Safety Oversight Audit Program discovered "modern training programs are not always in accordance with ICAO requirements." These variances include "(a) lack of consistency between international requirements and (good) industry practices . . . (and) training curricula (that does) not reflect modern training technologies and methodologies" (ICAO's Review of International Requirements for Flight Crew Licensing and Training 2006 ¶s 13 and14). Clearly, even within ICAO conforming States, training quality varies greatly. This is demonstrated even within Canadian flight training where two neighbouring flight schools can offer a drastically different quality of programming. As the Civil Aviation Daily Occurrence Reporting System displays, some flight training schools have a disregard for the Canadian Aviation Regulations, while other schools practice higher standards.

One solution to this dilemma could be the use of simulation. Simulator scenarios can be reproduced across the globe, providing consistency in instruction to students and trainers. Flight simulation is at the point where the learning experience can be equal to that of actual flight time on the aircraft, and in some circumstances more beneficial. Al Wardle, a pilot with over forty-five years of experience and over 13,000 flying hours, explains, "when I first got checked out on the Hercules (aircraft), we spent a week flying in the circuit. Now with the use of a simulator we spend 2 days in the air and the rest in the simulator. This essentially, replaces 60% of our airtime with the simulator, which is great because you can reproduce exact situations, pause or repeat exercises. Simulators are cost effective and efficient for both operators and pilots" (Wardle Personal Communication 20 March 2009). As mentioned, the benefit of simulation is within its efficiency. Since weather conditions become irrelevant to the commencement of training sessions, flight simulation offers efficiency. Furthermore, it is cost-effective with virtually zero maintenance. There is an opportunity to solve a majority of the industry's problems through a consistent training regime and rigorous, global, and standardized testing or accreditation. Training facilities themselves need to be monitored or spot-checked to ensure standards are continually being met. Training facilities would be at risk of losing their accreditation unless they pass routine scrutiny by ICAO. While the delay of creating a consistent training protocol may take a generation of pilots, the benefits would include consistency that the industry has yet to witness, regardless of where the training was received, therefore reducing inefficiencies and accidents.

Regulation

Aviation is one of the most highly regulated industries in the world, as well as one of the most complicated. ICAO has done an incredible job of standardizing regulations and standard operating procedures for all conforming states. ICAO's intentions are to create consistency, safety, and efficiency across the globe. However, transportation ministers of each State are not obligated to follow ICAO practices. In other words, the Minister of Transportation for Canada is going to do what is best for Canada, and the same can be said for any other country. Contradicting ICAO is not an issue taken lightly, but it is practiced. The United States Federal Aviation Administration (FAA) clearly describes its process for accepting ICAO standards and recommended practices (SARPs). The FAA states its three options regarding SARPs:

1. To adopt the proposed or revised SARP

2. To reject the proposed or revised SARP

3. To commit to adopt partial compliance (Assessing Compliance with ICAO 2007)

If the FAA decides to differ from the ICAO SARP (the latter two options), they have to file the differences with ICAO (Assessing Compliance with ICAO 2007). This is the problem: ICAO created the SARP because it was best for everyone in the aviation industry. However, States can reject the SARP on the basis that it does not suit their interests. An excellent example of how this can complicate issues is the ICAO recommendation for 406 emergency locator transmitters (ELT). While Canada has accepted the 406 ELT amendment, the United States has not (Psutka 2008). This creates a serious difference between neighbouring countries on a critical issue. ELTs are critical to locating and rescuing a pilot in a timely manner after a crash, but this efficiency may have been compromised. For example, if an American plane wanders into Canadian airspace, which is not uncommon, and crashes without one of the 406 ELTs, it complicates search and rescue activities. Canadian Owners and Pilot Association's President and CEO Kevin Psutka explains his concern regarding this SARP ratification, "the Canada Border Services Agency . . . processed over 63,000 foreign private aircraft in a one-year period from May 2007 to May 2008 and about 90% were US registered" (Psutka 2008 ¶10). Beyond the fact that this affects both country's economies negatively, safety is compromised due to the fact that the FAA decided not to comply with an ICAO recommendation. This example demonstrates why ICAO and the aviation industry as a whole would benefit from true, unequivocal, regulatory powers.

Establishing that ICAO should have actual regulatory powers is rather clear-cut; deciphering how the logistics could work is very difficult. To set precedent, the European Union (EU) will be used as an example. The EU represents twenty-seven countries in an economic and political partnership (Panorama of the European Union). The EU is broken down into three main groups, the European Parliament which represents the people, the Council of the European Union which represents national governments, and the European Commission which represents common EU interests (Panorama of the European Union). The system is democratic in nature. This arrangement is unique; voting takes place for the European Parliament across country borders. A similar system could be created for aviation with ICAO as the foundation. It can be argued that a certain amount of sovereignty is sacrificed if such a system were put in place. This would contradict article 1 of the Chicago Convention, ratified by ICAO, "the contracting states recognize that every state has complete and exclusive sovereignty over the airspace above its territory" (Convention on International Civil Aviation 1944 1). The reason for this article in the Chicago Convention is sound. Since the aviation industry is a crucial component of a nation's economy, there is a possibility that countries could use this process to damage enemies. A good example of this would be a number of third-world States (Afghanistan, Iraq, etc.) teaming up against America. This argument is the biggest obstacle before a "regulatory ICAO" could exist. However, just because the world is not in an appropriate climate to accept such a system presently does not mean the future does not hold promise.

The aviation industry has been a leader in the cooperation efforts of nations. It can be argued that ICAO is the United Nation's most successful organization. A possible solution to creating a fair system could include voting allocation based on aviation activity, rather than the current system; a single vote per State. The new system would have to be democratic with a quasi-judicial structure. As new markets emerge and growth occurs, conflicting regulations will cause more safety concerns. The EU provides the best example of how a beneficial relationship, with a small sacrifice in ultimate sovereignty, can exist. If aviation had an enforceable regulatory governing body, consistency and therefore safety would persist.

Conclusion

As the aviation industry expands, greater cooperation and consistency will be required. The inconsistencies in the current global aviation industry are only going to be exaggerated as growth occurs, including differences in language standards, varying cultures, training, and regulatory powers. The proper use of language will become increasingly important as more foreign pilots enter the world of aviation. Concise and consistent phraseology must be employed. Varying norms and national cultural views should be substituted for a safety culture while flying or controlling aircraft. The easiest way to implement a number of these changes is through training. The next generation of aviators should be trained to a recognized, global standard. ICAO has the possibility to be a very effective governing body for the aviation industry. Although current conditions do not allow ICAO to enforce practices, the future may hold the opportunity for such a system. Consistency creates an environment with safety in mind; surprises are not sought after or welcome in aviation. With growth in air travel, the impacts of cultural and national differences will become exaggerated.

Future Impacts of Automation on the Aviation Industry

By Kenneth Polonenko

The aviation industry is an ever-changing entity. Aircraft have evolved from simple, single-engine crafts with limited range and endurance to incredibly complex, multi-engine machines capable of transporting hundreds of people halfway around the globe. As technology enhances the aircraft's ability, new ways of fostering that capability are required to ensure safe and efficient transportation. The flight deck is the location where required information is provided to the pilots for decision-making purposes and the place where control of the aircraft is accomplished. Cockpit displays have been the subject of intense psychological and technological research, and, as a result, the trend is towards possessing significantly enhanced technical capability, complexity, and sophistication (Hettinger and Haas 2000).

Aircraft have become increasingly automated, changing the role of pilots considerably. Automation can be defined as "a machine agent or agents carrying out tasks (or partial tasks) once allocated to the human" (Hoeft, Kochan, and Jentsch 2006 244). There is a recurring theme that the human will eventually be removed from the cockpit team altogether, with aircraft having the ability to safely fly itself gate-to-gate. This paper will present the current and emerging automation technologies and argue that the human will not be removed from the cockpit within the next ten years, with a focus becoming more on training for the automated environment. Although there is great advancement in automation in the general aviation and military realm, this paper's focus is within the civil aviation, air carrier field.

History of Automation

The earliest aircraft were operated by a single pilot, due to the simplicity, size, and elementary nature of the planes and their systems. As the size, range, and sophistication of the equipment developed, a second pilot was added to the cockpit to assist the captain in controlling the aircraft and monitoring the systems. In the 1930s, air-cooled engines replaced water-cooled engines, which reduced weight and allowed for larger and faster aircraft. The instrumentation in the cockpit also advanced, with more reliable turn coordinators, altimeters, vertical speed indicators, airspeed indicators, compasses, and the introduction of the artificial horizon (Koop 1998). With more trusted instruments, the autopilot was able to flourish in the 1930s. Originally invented in 1908 by Lawrence Sperry, first-generation autopilots became increasingly present in the cockpit in the 1930s. These systems, such as the Sperry A-12 gyropilot, were originally developed to improve aircraft stability and lower passenger sickness (Hoeft, Kochan, and Jentsch 2006). The earliest autopilots were single-axis, only able to control roll using four gyroscopes (including the attitude indicator and heading indicator) and a simple feedback loop, hydraulically operating the elevators and rudder.

With the technological advancement of aircraft in the 1950s, most autopilots in transport aircraft had three-axis capability (controlling roll, pitch, and yaw). The main argument for the rapid development and installation of automation was concern for human performance limitations as the aircraft were flying higher, faster, and farther (Hoeft, Kochan, and Jentsch 2006).

Studies focused on pilot stress and fatigue, but in the 1960s, aircraft manufacturers realized the economic savings that could be realized by replacing the flight engineer, navigator, and radio operator with the new technology. Computer software and electronic sensors became essential within the autopilot feedback loop, controlling hydraulically actuated servos that manipulated the control surfaces as needed (Osdner and Hutchingson 1961).

Current Automation

The technology in the cockpit has developed into a complex system that has the ability to automatically control every aspect of the aircraft's flight in terms of lateral and vertical navigation (LNAV and VNAV) and speed from immediately after takeoff to the end of the landing roll (Harris 2004). This is accomplished through the inputs for both internal and external navigational sources and the management of the engine thrust to maintain optimum power at optimum economy for the appropriate phase of flight. In modern transport aircraft, this is achieved through the Flight Management System (FMS), the "brain" of the airplane, which is an integration of four major systems: the flight management computer system (FMCS), the digital flight control system (DFCS), the auto-throttle (A/T), and the inertial reference system (IRS) (Wells 2001). The purpose of this computerized system is to reduce pilot workload and to achieve the best possible fuel economy. It allows crews to have access to the performance, navigational, and system status at any time. The FMS can be interfaced with the automatic flight control and A/T systems so that the aircraft can be fully automatic for all phases of flight, requiring very little direct pilot input. The FMS also provides information to the flight deck displays such as the Electronic Flight Instrument System (EFIS), Engine Indicating and Crew Alerting System (EICAS), and Engine Centralized Aircraft Monitoring (ECAM), which are all computer-generated displays to provide the pilot with essential information (Jukes 2004). It has become impossible for crew members to monitor and control the automation process using the traditional instruments, so these tools were developed to ensure safety.

The actual automatic flight system can be utilized to perform different tasks. In the simplest state, it can hold a preselected heading and altitude. It can also track navigational aids, maintain a selected vertical speed, maintain an indicated speed or mach value, fly an approach, and, if rated, land the aircraft. In order to execute Category 3b or 3c ILS approaches (auto-land and auto-rollout), certain requirements must be met, not only within the aircraft equipment, but with ground-based equipment, as well as crew training. For example, the aircraft must be equipped with a minimum of two, fully independent Flight Directors (FDs), Air Data Computers (ADC), radio altimeters, primary flight displays (PDFs), generators, VHF Nav and Com receivers, Pitot-Static and related probes and heaters, and any additional related equipment as certified for the respective aircraft, and all must meet rigorous safety requirements. In general, pilots may use automation for both tactical and strategic control of the aircraft; tactical focusing on the short term (maintain a certain heading), while strategic is affected on a longer time horizon (for example, land on a certain runway following a specific approach) (Zushlag 2002).

The automated cockpit has been the driving force behind the safe, accurate, and economical operations of the airline industries that has permitted the increase of capacity throughout the past fifty years, along with the corresponding rise in safety and efficiency (Olson 2001).

Future Automation and Cockpit Enhancements

The rate of change of automation will be slower than the rate of change of other technological advancements. As the technology increases, the cockpit will change accordingly, but the amount of automation will not change for many years. For example, the introduction of the Global Positioning System (GPS) "is the single most important event that will make today's air traffic system obsolete" (Wells 2001 192). The flight deck and avionics will change appropriately to accommodate the GPS-based Wide Area Augmentation System/Local Area Augmentation System (WAAS/LAAS) approaches and landings, Global Navigation Satellite System

(GNSS)-based en route navigation, and automatic dependent surveillance (Wells 2001). Also, tools such as Ground Proximity Warning Systems (GPWS), Heads Up Displays (HUDs) and synthetic vision will become increasingly standard in cockpits, greatly influencing the reduction of Controlled Flight Into Terrain (CFIT), but not automation.

Also, the concept of "free flight" is being regarded as the air transport management system of the future and represents one of the greatest changes in the future of aviation. Currently, airlines lose billions of dollars a year due to the inefficient ATC environment (Wells 2001). The increased number of traffic delays is an indication that the capacity of the existing air traffic system is reaching its limit. The relation between traffic density and delays is not linear, and a small increase in density results in an exponential increase of delays (Theunissen 2007). Future systems must be able to allow for greater capacity, minimizing delays at the same time, accomplished in free flight (beginning to be referred to as "NextGen") by direct routes flown by aircraft at their optimum altitude. Aircraft will no longer be restricted to flying along certain airways and at regulatory altitudes. In order to maintain safe aircraft spacing, more advanced Traffic Collision Avoidance System (TCAS) systems will be used, and ADS-B transponders will replace Mode S transponders, allowing an aircraft to transmit its location and intent to every other aircraft in the vicinity. Automation will be used to allow the aircraft to determine the most efficient route and altitude, and also identify potential collisions and present solutions.

Currently, the majority of air transport aircraft are flown by autopilot several seconds after takeoff (usually 400–600 feet above ground level) until several moments before landing, as per manufacturer limits. In order for automation to increase, taxiing, takeoff, landing, and docking must be included, essentially removing the pilot from all direct control operations. It has been widely debated whether or not pilots will eventually be completely removed from the airline cockpit altogether, allowing the aircraft to fully fly itself. Change will occur only when the airlines believe that there is definite economic benefit in doing so. There has been great use of Unmanned Air Vehicles (UAVs) in the military realm, with the air forces taking advantage of removing the pilot from hostile environments and the weight savings from removing the needed equipment for human-machine interface (Kochan 2005). These advantages do not apply significantly to civil aviation. In fact, regarding safety, removing the pilot can have several disadvantages. The onboard computer may not have the knowledge or ability to resolve an improbable malfunction that has occurred, while an onboard pilot can manage the problem effectively. Dr. Theunissen of the Delft University of Technology summarized the safety advantages of maintaining pilots in the control loop as: "Humans can respond perceptually to a changing environment and to relations in the environment. They can go beyond the information immediately given, respond to low-probability occurrences, and adopt alternative strategies and alternate modes of performance when necessary. In short, humans are flexible" (Theunissen 2007).

An ideal example demonstrating the flexibility of pilots over automation is the US Airways Flight 1549 that completed an emergency landing in the Hudson River 5 February 2009. The pilots were required to make an instantaneous decision to determine the appropriate course of action after losing thrust in both engines immediately after takeoff due to bird strikes. Without the human element of past experience, knowledge, and ability to hand-fly the Airbus aircraft, the situation could have arguably ended with hundreds of lives lost. In order for automation to be effectively utilized, information must be programmed into the computers and verified by the pilots. Automation cannot currently preprogram itself, and human intervention is needed when circumstances change, as in the case of US Airways 1549.

Even though removing pilots from airline operations is unlikely in the near future, research is being done to explore the possibility. For example, Innovative Future Air Transport System (IFATS) is a European Commissioned project studying the concept of a fully automated air transport system, replacing both pilots and controllers with ground operators. With a vision of the long-run (2040), IFATS can be expected to reduce the environmental impacts of air travel by increasing efficient traffic flow, allow variations of the airport approach fight path, improve the security/safety of ATS by automated collision avoidance, and eliminating the possibility of hijacks, and reduce operator's cost by eliminating the need for pilots (Innovative Future Air Transport System 2004).

Although it is difficult to predict the future of the airliner flight deck, one can be certain the human component will not be removed until the airliners and flying public deem the advantages to be greater than the disadvantages. When, and if, these changes occur, they will happen in small increments rather than giant steps. In a similar example, the replacement of traditional navigational aids such as NDBs, VORs, and ILSs with satellite-based systems has been a slow process, requiring testing, trust gathering, regulation generating, and implementation. A comparable process would be required for full automation. Airlines would incur an enormous expense in retrofitting and updating their current fleet of aircraft, a progression that could take upwards of twenty years. Public perception will also dictate the rate at which automation succeeds in airliners. The traveling public expects trained and experienced pilots to command the aircraft, and air travel demand may significantly drop if humans are no longer in control.

Also, by removing the human from the cockpit, a significant portion of the workforce would be eliminated; the pilot group, regulatory body that governs pilots, and those who train the pilots. These groups are generally located within a higher tax bracket, and the government would lose a significant amount of tax income if these jobs disappeared.

Automation will continue to be a prevalent theme in the modern aircraft. New technology will be introduced to enhance the automation, efficiency and safety, but humans will continue to play a vital role for the near future. Automation has changed the role of the pilot dramatically, shifting it from direct operational control of the aircraft to managing the systems that control the aircraft. Even though great advances have been made, there will always be a clash when humans and machines attempt to interact.

Human Factor Problems with Automation

Even though automation has been present in airliners for upwards of seventy years and many of the safety-critical functions have been transferred away from human control, there is the potential for new classes of error (Zuschlag 2002). A famous example of the struggle between human and autopilot is the crash of an Airbus A300-600 on approach in Nagoya, Japan, in 1994.

The first officer inadvertently engaged the "go-around" function of the autopilot, causing the aircraft to attempt to fly away from the ground using the pitch-trim system. Unaware of the pitch problem, the pilots struggled, continuing to push the nose down, causing further opposite action from the pitch-trim system. The aircraft eventually took a vertical nose-up position, stalled, and crashed. A recent example is the 12 February 2009 crash of a Dash 8-400 near Buffalo, N.Y. Although the NTSB has not released the final results, preliminary indications suggest the pilots were unaware of the significant ice build up because of the use of autopilot, which led to an unrecoverable stall several hundred feet above the ground.

One particular problem with the interaction between humans and automation is "automation bias," the practice of humans treating information given by an automated system as fact without seeking clarification or confirmation (Hoeft, Kochan, and Jentsch 2006). In a 2007 study to develop a comprehensive list of flight automation issues that can be used for further research and in design manufacture, Ken Funk and his colleagues at the Oregon State University generated ninety-four possible problems between the human and automation. The five most prevailing issues as determined by the study include:

- Understanding of automation may be inadequate.

- Behaviour of automation may not be apparent.

- Pilots may be overconfident in automation.

- Displays (visual and aural) may be poorly designed.

- Training may be inadequate (Funk 2007).

It can be argued that the first three issues listed above can stem from the fifth, that the training may be inadequate. As automation continues to be a predominant tool in the current and future cockpit, humans must be trained and technology developed in order to work as a team.

The concept of automation systems being part of the cockpit "team" developed as early as World War II, when pilots nicknamed their autopilots "George" (Hoeft, Kochan, and Jentsch 2006). Crews must be trained to use automation to such an expert level that common questions found in the cockpit are eliminated, such as: "What is it doing now?" "Why is it doing that?" and "What will it do next?" Although design faults such as inconsistent interfaces, operating mode proliferation, system insulation, and distributed controls and displays play a critical role in understanding the human factor difficulties involved with automation (Zushclag 2002), technological advances that will evolve in the future can largely eliminate this factor. Only training can allow crews to work effectively and efficiently as a team.

Conclusion

Automation has developed into a tool used in the airline industry as a means to improve safety, efficiency, and pilot workload. Automation will continue to be prevalent and widely used in the future, but the human pilot will remain a vital component of the flight deck team. Humans are still required to program and monitor the automation, and for when unknown or unpredictable malfunctions occur, using innovation and unprogrammable knowledge to guide the aircraft to safety. Technology will continue to develop tools in aiding automation and the pilots, creating an increasingly safer and efficient operation. There are many problems when humans attempt to work alongside automation, both with design intolerances and with how the human interacts with the machine. In the future, the interface between the human and machine will be better suited for safer airline transportation, and the pilots will be better trained to include the automation as a team member, all working cohesively towards the same objective—safe and efficient flight.

PRACTICE ACTIVITY

What technology, practice, or innovation do you think will have the greatest impact on the future of aviation? Describe your ideas below.

_____

_____

_____

_____

_____

_____

_____

_____

_____

_____

_____

_____

_____

_____

_____

_____

_____

_____

_____

If you are interested in contributing a student essay to a future edition of *Canadian Aviation,* submit your 2000–2500-word paper, answering the question above, to the author.

Resources

Chapter 1

Abbott, Patrick. *Airship: The Story of the R.34 and the First East-West Crossing of the Atlantic by Air.* New York: Charles Scribner's Sons, 1973.

Bashow, David L. *All the Fine Young Eagles: In the Cockpit with Canada's Second World War Fighter Pilots.* Toronto: Stoddart, 1996.

———. *No Prouder Place: Canadians and the Bomber Command Experience, 1939–1945.* St. Catharines, ON: Vanwell, 2008.

Bélanger, Réné. *Conquering the North Shore by Air: Development of Aviation on the North Shore, 1919–1954.* Quebec: Éditions La Liberté, 1978.

Corley-Smith, Peter. *Barnstorming to Bush Flying: British Columbia's Aviation Pioneers, 1910–1930.* Victoria: Sono Nis Press, 1989.

Corn, Joseph J. *The Winged Gospel: America's Romance with Aviation, 1900–1950.* New York: Oxford University Press, 1983.

Ellis, Frank. *Canada's Flying Heritage.* Toronto: University of Toronto Press, 1954.

Fritzsche, Peter. *A Nation of Fliers: German Aviation and the Popular Imagination.* Cambridge: Harvard University Press, 1992.

Fuller, G.A., J.A. Griffin, and K.M. Molson. *125 Years of Canadian Aeronautics: A Chronology, 1840–1965.* Willowdale, ON: Canadian Aviation Historical Society, 1983.

Goldstein, Laurence. *The Flying Machine and Modern Literature.* Bloomington: Indiana University Press, 1986.

Greenhous, Brereton, et al, *The Crucible of War, 1939-1945.* Toronto: University of Toronto Press, 1994.

Jackson, Robert. *Airships: A Popular History of Dirigibles, Zeppelins, Blimps, and Other Lighter-Than-Air Craft.* Garden City, N.Y.: Doubleday, 1973.

Myers, Patricia A. *Sky Riders: An Illustrated History of Aviation in Alberta, 1906–1945.* Saskatoon: Fifth House, 1995.

Pigott, Peter. *Flying Colours: A History of Commercial Aviation in Canada.* Vancouver: Douglas & McIntyre, 1997.

———. *Wingwalkers: The Story of Canadian Airlines International.* Vancouver: Harbour Publishing, 2002.

Ralph, Wayne. *Barker VC: William Barker, Canada's Most Decorated War Hero.* Toronto: Doubleday, 1997.

Reid, Sheila. *Wings of a Hero: Canadian Pioneer Flying Ace Wilfrid "Wop" May.* St. Catharines: Vanwell, 1997.

Smith, Philip. *It Seems Like Only Yesterday: Air Canada, the First 50 Years.* Toronto: McClelland and Stewart, 1986.

Vance, Jonathan F. *High Flight: Aviation and the Canadian Imagination.* Toronto: Penguin Canada, 2002.

West, Bruce. *The Firebirds: How Bush Flying Won Its Wings.* Toronto: Ministry of Natural Resources, 1974.

Wise, S.F. *Canadian Airmen and the First World War.* Toronto: University of Toronto Press, 1980.

Chapter 2

ADF. *Job Description.* Washington, D.C.: Airline Dispatchers Federation, 2009. www.dispatcher.org/index.php?option=com_content&task=view&id=44&Itemid=82 (accessed 30 April 2009).

Alexander, K.L. "Airlines find fuel prices tough to swallow," *Washington Post* (3 June 2004): E03.

CAMC. *Follow-up to the human resources study of pilots in Canada.* Ottawa: Canadian Aviation Maintenance Council, 2003.

Careless, J. "Fractional ownership ready for takeoff" *Wings Magazine* (n.d.). http://www.wingsmagazine.com/content/view/431/ (accessed 17 April 2009).

COPA. "Canadian general aviation: Past, present, & future." Ottawa: Canadian Owners and Pilots Association, 2008. http://www.copanational.org/non-members/index.htm (accessed 22 March 2009).

Cribb, R., F. Vallance-Jones, and T. McMahon. "Jetsgo problems ignored" (16 June 2006). thestar.com: http://www.thestar.com/News/article/144218 (accessed 3 May 2009).

Morris, D "De-iceman cometh," *Weatherwise* (27 March 2009).

"Most admired corporates: #9 WestJet," *Financial Post* (21 January 2009). http://www.financialpost.com/careers/story.html?id=1200493 (accessed 2 May 2009).

Strait, B. "*Welcome to very light jet magazine*," *Very Light Jet Magazine* (2006). http://www.verylightjetmagazine .com/editorial.html (accessed 4 October 2008).

Transport Canada. "Canadian Civil Aircraft Register," *Transport Canada Civil Aviation* (1 January 2009). http://www.tc.gc.ca/aviation/activepages/ccarcs/aspscripts/en/monthsumairbycatresult.asp (accessed 2 May 2009).

———. "Summary of flight crew and air traffic control licenses," *Civil Aviation* (8 January 2009). http://www.tc.gc.ca/civilaviation/general/personnel/stats/stats007.htm (accessed 14 April 2009).

———. "Flight Training Units," *Flight Training* (2009). http://www.tc.gc.ca/aviation/activepages/ftae/Index.aspx (accessed 16 April 2009).

———. *Civil Aircraft Activity in Canada.* Ottawa: Transport Canada, 2002.

Chapter 3

Air Force Historical Studies Office. *High Flight.* Washington, D.C.: Air Force Historical Studies Office, n.d. http://www.airforcehistory.hq.af.mil/PopTopics/highflight.htm (accessed 11 May 2009).

Canada's Air Force. "Air Force Wings Across Canada," National Defence (1 October 2006). http://www .airforce.forces.gc.ca/site/orgdocs/organization3_e.asp (accessed 10 May 2009).

———. "CC-177 Globemaster III," *Aircraft* (1 August 2007). http://www.airforce.forces.gc.ca/site/equip/ cc177/photos_e.asp (accessed 10 May 2009).

———. "Search and Rescue," *National Defence* (18 March 2007). http://www.airforce.forces.gc.ca/site/ athomedocs/athome_2_e.asp (accessed 11 May 2009).

———. "Welcome to Air Force History," National Defence (9 December 2008). http://www.airforce.forces .gc.ca/v2/hst/index-eng.asp (accessed 12 April 2009).

CEFCOM. "CEFCOM Home: What We Do," National Defence and the Canadian Forces (6 May 2009). http://www.comfec-cefcom.forces.gc.ca/pa-ap/about-notre/index-eng.asp (accessed 12 May 2009).

Fetterly, R. "The cost of peacekeeping: Canada," *The Economics of Peace and Security Journal* (2006): 47–53.

Lee, R.M. *Death and Deliverance.* Toronto: Macfarlane Walter & Ross: 1992.

Lund, W. "Integration and Unification of the Canadian Forces," CFB Esquimalt Naval and History Museum (2007). http://www.navalandmilitarymuseum.org/resource_pages/controversies/unification.html (accessed 11 May 2009).

McLean, A. "New Canadian military drones will carry bombs," Canada.com. (7 March 2009) www.canada.com/Technology/Canadian+military+drones+will+carry+bombs/1364912.story.html (accessed 11 May 2009).

National Defence. "Canadian Forces Disaster Assistance Response Team," National Defence (10 January 2005). http://www.comfec-cefcom.forces.gc.ca/pa-ap/nr-sp/doc-eng.asp?id=301 (accessed 12 May 2009).

———. "Canadian Operational Support Command (CANOSCOM)," National Defence and the Canadian Forces (1 August 2006). http://www.canoscom-comsocan.forces.gc.ca/index-eng.asp (accessed 12 May 2009).

———. "CANSOFCOM Core Tasks," National Defence and the Canadian Forces (30 April 2008). http://www.cansofcom.forces.gc.ca/gi-ig/cct-tbc-eng.asp (accessed 12 May 2009).

———. (2009, April 28). *Welcome to Canada Command.* Retrieved, from National Defence: http://www.canadacom.forces.gc.ca/site/index-eng.asp (accessed 12 May 2009).

National Defence and the Canadian Forces. "Operations," National Defence (19 December 2008). http://www.forces.gc.ca/site/operations/current-ops-courante-eng.asp (accessed 9 May 2009).

National Search and Rescue Secretariat. "Review of SAR Response Services," Reports (February 17, 2004). http://www.nss.gc.ca/site/reports/responsereview_e.asp (accessed 11 May 2009).

National Search and Rescue Secretariat. "Search and Rescue," Government of Canada (21 April 2009). http://www.nss.gc.ca/site/index_e.asp (accessed 12 May 2009).

North American Aerospace Defense Command. "About Us," North American Aerospace Defense Command (n.d.). http://www.norad.mil/about/index.html (accessed 12 May 2009).

Royal Military College of Canada. "About RMC," National Defence (8 January 2009). http://www.rmc.ca/ about-apropos-eng.asp (accessed 11 May 2009).

United Nations Peacekeeping. "Overview," United Nations Peacekeeping (2008). http://www.un.org/Depts/
dpko/dpko/ (accessed 12 April 2009).

Veterans Affairs Canada. "Tomb of the Unknown Soldier," Veterans Affairs Canada (7 May 2005).
http://www.vac-acc.gc.ca/remembers/sub.cfm?source=memorials/tomb (accessed 11 May 2009).

Watt, A. "On Windswept Heights," Canada's Air Force (6 April 2009). http://www.airforce.forces.gc.ca/
v2/hst/page-eng.asp?id=884 (accessed 10 April 2009).

Chapter 4

ACI. *Airport cost approach comparison.* Washington, D.C.: Airports Council International, 2003.

CAC. "Reaching beyond for the community," Canadian Airports Council (2008).
http://www.cacairports.ca/english/canadas_airports/community.php (accessed 12 February 2009).

CATSA. "Canadian Air Transport Security Authority," About CATSA (5 June 2007). http://www.catsa.gc.ca/
english/about_propos/ (accessed 7 February 2009).

Facette, J. "Canadian Airports Council," Aviation standing committee presentation (13 May 2008).
http://www.cacairports.ca/Docs_2008/AviationStandingCommitteeApril2008.pdf (accessed 13 February
2009).

ICAO. "International Civil Aviation Organization Presentation by the United States of America," Meeting the
100 Percent Hold Baggage Screening Requirement in January 2006 (11–14 October 2005). http://www.icao
.int/icao/en/ro/nacc/meetings/2005/NACC_DCA2/nacc02ip07.pdf (accessed 9 February 2009).

Transport Canada. "National Airports Policy," Transport Canada (12 January 2009). http://www.tc.gc.ca/
programs/airports/policy/NASImplementation.htm (accessed 16 May 2009).

Wong, N. "Logan will install body scanners," *The Boston Globe* (1 July 2008). http://www.boston.com/
business/technology/articles/2008/07/01/logan_will_install_body_scanners/ (accessed 6 February 2009).

www.enviro.aero. "Noise" (2009). www.enviro.aero:http://www.enviro.aero/Noise.aspx (accessed 12 January
2009).

Chapter 5

Canada Gazette. *Part 1—Notices and proposed regulations.* Ottawa: Queen's Printer, 2009.

COSPAS SARSAT. "Cospas–Sarsat Phase-Out of 121.5/243 MHz Alerting Services," International Satellite
System for Search and Rescue (October 2008). http://www.cospas-sarsat.org/FirstPage/121.5PhaseOut.htm
(accessed 6 May 2009).

Government of Canada. "Regulations," Cabinet Directive on Streamlining Regulation (2007, 1 April).
http://www.regulation.gc.ca/directive/directive00-eng.asp (accessed 24 April 2009).

ICAO. *Convention on International Civil Aviation.* Montreal: International Civil Aviation Organization, 2006.

———. "International Civil Aviation Organization," Memorandum on ICAO (n.d.). http://www.icao.int/
icao/en/pub/memo.pdf (accessed 20 April 2009).

———. "Strategic Objectives of ICAO for 2005–2010," International Civil Aviation Organization
(17 December 2004). http://www.icao.int/icao/en/strategic_objectives_2005_2010_en.pdf (accessed
21 April 2009).

Minister of Civil Aviation Tribunal. *Civial Aviation Tribunal of Canada.* Ottawa: Canada Communications
Group—Publishing, 1997.

Transport Canada "About the Canadian Aviation Regulations (CARs)," Transport Canada Civil Aviation
(31 December 2007). http://www.tc.gc.ca/civilaviation/regserv/affairs/cars/about.htm (accessed 23 April
2009).

———. "CARAC Management Charter and Procedures," Transport Canada—Civil Aviation: (April 2008).
http://www.tc.gc.ca/civilaviation/RegServ/Affairs/carac/Charter/menu.htm#DIVI_2.0 (accessed 23 April
2009).

———. "Civil Aviation," Policy and Regulation Services (31 December 2007). http://www.tc.gc.ca/
civilaviation/RegServ/affairs/carac/about.htm (accessed 20 April 2009).

———. "TP 13918—406 MHz ELT . . . The Next Generation of ELT," Transport Canada—Civil Aviation
(9 July 2008). http://www.tc.gc.ca/CivilAviation/SystemSafety/Videos/tp13918Transcript.htm (accessed
23 April 2009).

Chapter 6

Aviation Safety Letter. *Safety Target—Runway Incursions.* Ottawa: Transport Canada, 2002.

Nav Canada. "A General Overview of the Company," Nav Canada (December 2008). http://www.navcanada .ca/NavCanada.asp?Language=en&Content=ContentDefinitionFiles\Newsroom\Backgrounders\ nc_glance.xml (accessed 27 January 2009).

———. "Facilities," About us (n.d.). http://www.navcanada.ca/NavCanada.asp?Language=EN&Content= ContentDefinitionFiles%5CAboutUs%5CWhatWeDo%5CFacilities%5Cdefault.xml#acc (accessed 2 February 2009).

NTSB. *Aircraft Accident Report NTSB/AAR-87/07.* Washington, D.C.: National Transportation Safety Board, 1986.

Transport Canada "National Airports Policy—Airport Programs," Transport Canada—Air Transportation (12 January 2009). http://www.tc.gc.ca/programs/airports/policy/NASImplementation.htm (accessed 26 January 2009).

Chapter 7

ASRS. "Report Number 611329," Aviation Safety Reporting System (March 2004). http://www.asias.faa.gov/ pls/portal/STAGE.ASRS_BRIEF_REPORT?RPT_NBR=611329&AC_VAR=TRUE&RPRT_VAR= TRUE&ANMLY_VAR=TRUE&SYN_VAR=TRUE&NARR_VAR=TRUE&NARR_SRCH=1%20fell%20asleep (accessed 8 January 2009).

Bourgeois-Bougrine, S., P. Carbon, C. Gounelle, R. Mollard, and A. Coblentz. "Perceived Fatigue for Short- and Long-Haul Flights: A Survey of 739 Airline Pilots," *Aviation, Space, and Environmental Medicine* (2003): 1072–1077.

Caldwell, J.A., K.K. Hall, and B.S. Erickson. "EEG Data Collected from Helicopter Pilots in Flight Are Sufficiently Sensitive to Detect Increased Fatigue from Sleep Deprivation, *International Journal of Aviation Psychology* (2002): 19–32.

Casey, S. *Set phasers on stun: And other true tales of design, technology, and human error.* Santa Barbara, CA: Aegean, 1998.

Committee on Transportation and Infrastructure. *FAA Oversight of Falsifications on Airman Medical Certificate Applications.* Washington, D.C.: U.S. House of Representatives: 2007.

Endsley, M.R. "Measurement of situation awareness in dynamic systems," *Human Factors* 37 (1995): 65–84.

FITS. "Flight instructor training module," Vol. 1, FAA/Industry Training Standards. Flight Standards Service. General Aviation and Commercial Division (AFS-800). Washington, D.C.: Federal Aviation Administration (n.d.). http://www.faa.gov/education_research/training/fits/training/flight_instructor/media/Volume1.pdf (accessed 10 September 2008).

Flin, R., P. O'Connor, and K. Mearns. "Crew resource management: Improving team work in high reliability industries," *Team Performance Management,* 8.3/4 (2002): 68–78.

Helmreich, R.L., A.C. Merritt, and J.A. Wilhelm. "The evolution of crew resource management training in commercial aviation," *The International Journal of Aviation Psychology,* 9.1, (1999): 19–32.

Jensen, R.S. *Pilot judgement and crew resource management.* Burlington, Vt: Ashgate, 1995.

Kearns, S.K. "The effectiveness of guided mental practice in a computer-based single pilot resource management training program," Proquest Digital Dissertations Database. UMI No. 3288814 (2007).

Mann, M.B. "NASA Statement on Pilot Fatigue," Hearing on Pilot Fatigue (1999, August 3). http://www.hq.nasa.gov/office/legaff/mann8-3.html (accessed January 7, 2009).

NBAA Safety Committee. "VLJ training guidelines for single pilot operations of very light jets and technical- ly advanced aircraft," National Business Aviation Association (2005). http://web.nbaa.org/public/ops/ safety/vlj/VLJ_Training_Guidelines.pdf (accessed July 18, 2008).

Nguyen, L. "Co-pilot's illness diverts flight," *Calgary Herald* (2008, January 30).

NTSB. *Aircraft Accident Report: Eastern Air Lines, Inc.* Washington, D.C.: National Transportation Safety Board, 1973.

———. *General aviation accidents involving visual flight rules flight into instrument meteorological conditions.* Washington, D.C.: Author, 1989.

———. *Aircraft accident report: United airlines flight 232.* Washington, D.C.: National Transportation Safety Board, 1990.

Proctor, M.D., M. Panko, and S. Donovan. "Considerations for training team situation awareness and task performance through PC–gamer simulated multi-ship helicopter operations," *International Journal of Aviation Psychology,* 14.2 (2004): 191–205.

Reinhart, R.O. *Basic Flight Physiology.* New York: McGraw Hill, (1996).

Salas, E., C.S. Burke, C. Bowers, and K.A. Wilson. "Team training in the skies: Does crew resource management (CRM) training work?" *Human Factors,* 43.4, (2001).: 641–674.

Spitzer, R.L., M. Terman, J.B. Williams, J.S. Terman, U.F. Malt, F. Singer, et al. "Jet Lag: Clinical Features, Validation of a New Syndrome-Specific Scale, and Lack of Response to Melatonin in a Randomized, Double-Blind Trial," *American Journal of Psychiatry* (2008).:1392–1396.

Strait, B. "Welcome to very light jet magazine," *Very Light Jet Magazine* (2006). http://www.verylightjetmagazine.com/editorial.html (accessed October 4, 2008).

Transport Canada. "Part VI—General Operating and Flight Rules," Canadian Aviation Regulations (2008, December 30). http://www.tc.gc.ca/CivilAviation/Regserv/Affairs/cars/PART6/602.htm (accessed January 6, 2009).

Trollip, S.R., and R. S. Jensen. *Human factors for general aviation.* Englewood, CO.: Jeppesen Sanderson, 1991.

Wickens, C.D., J.D. Lee, S.E. Gordon, and Y. Liu. *An introduction to human factors engineering.* New York: Longman, 2003.

Chapter 8

BBC. "Bones confirm Steve Fossett death," BBC News (3 November 2008). http://news.bbc.co.uk/2/hi/americas/7707397.stm. (accessed 11 January 2009).

Chapter 9

Flight Safety Foundation. "Aviation grapples with human-factors accidents," Flight Safety Digest (May 1999). http://www.flightsafety.org/fsd/fsd_may99.pdf (accessed 11 January 2009).

National Inventor's Hall of Fame Foundation. "Inventor's Profile—C. Donald Bateman," Hall of Fame (2002). http://www.invent.org/hall_of_fame/231.html (accessed 27 January 2009).

NTSB Aviation. "Cockpit voice recorders (CVRs) and flight data recorders (FDRs)," National Transportation Safety Board—Aviation (September 2004). http://www.ntsb.gov/aviation/CVR_FDR.htm (accessed 15 January 2009).

Walsh, N.P., and L. Harding. "Nothing left to lose: grief-crazed murder suspect haunted by family's air deaths," Guardian.co.uk World News: (28 February 2004). http://www.guardian.co.uk/world/2004/feb/28/russia.lukeharding (accessed 15 January 2009).

Chapter 10

Hawkins, F. H. *Human factors in flight.* Aldershot, UK: Ashgate, 1987.

ICAO. "ICAO SMS Module 03—Introduction to safety management," Safety Management: (15 November 2008). http://www.icao.int/anb/safetymanagement/training/training.html (accessed 19 March 2009).

ICAO. "Safety Management Manual," Second Edition, Doc 9859, International Civil Aviation Organization (2008). http://www.icao.int/anb/safetyManagement/DOC_9859_FULL_EN.pdf (accessed 19 March 2009).

Moshansky, V.P. *Commission of inquiry into the Air Ontario crash at Dryden, Ontario—Final Report.* Toronto: Minister of Supply and Services Canada 1992.

Reason, J. *Human Error.* New York: Cambridge University Press, 1990.

Transport Canada. *Moving Forward: Changing the Safety and Security Culture.* Ottawa: Transport Canada: 2007.

———. *Risk management and decision making in civil aviation. Type 2A, short process.* 4th Ed. Ottawa: Transport Canada: 2004.

Chapter 11
Synthetic and Enhanced Vision Systems

Arthur, J., L. Kramer, and R. Bailey Flight test comparison between enhanced vision (FLIR) and synthetic vision systems. *Proceedings of SPIE, 5802, 25–36.* doi:10.1117/12.604363. (2005).

Arthur, J., L. Prinzel, L. Kramer, R. Bailey, and R. Parrish. "CFIT prevention using synthetic vision," *Proceedings of SPIE, 5081, 146–157.* doi:10.1117/12.487291 (2003).

Arthur, J., L. Prinzel, K. Shelton, L. Kramer, S. Williams, and R. Bailey. "Design and testing of an unlimited field-of-regard synthetic vision head-worn display for commercial aircraft surface operations," *Proceedings of SPIE, 6559, 65590E1–19.* doi:10.1117/12.719695 (2007).

Associated Press, The. "Honeywell predicts bright future for synthetic vision in business jets" (2008). http://www.iht.com/articles/ap/2008/05/22/business/ EU-FIN-COM-Switzerland-Honeywell-Jet-Vision.php (accessed 8 March 2009).

Boeing Aircraft Inc. "Statistical Summary of Commercial Jet Airplane Accidents: Worldwide Operations 1959–2007" (2008). http://www.boeing. com/news/techissues/pdf/statsum.pdf (accessed 3 March 2009).

French, G., D. Murphy, and W. Ercoline. "Flare cue symbology and EVS for zero-zero weather landing," *Proceedings of SPIE, 6226, 62260L1–11.* doi:10.1117/12.666601 (2006).

Helicopter Association International. "U.S. Helicopter 10-Year Accident Summary Statistics 1997–2006" (2007). http://www.rotor.com/portals/12/Statistics%20Summary%201997-2006.pdf (accessed 5 March 2009).

Hemm, R. & Houser, S. (2001). *NASA Contractor Report: A Synthetic Vision Preliminary Integrated Safety Analysis.* Retrieved March 1, 2009, from http://gltrs.grc.nasa.gov/ reports/2001/CR-2001-211285.pdf.

Honeywell International (2009). *Technology Feature: Bad Visibility Not a Problem with Synthetic Vision System.* Retrieved March 3, 2009, from http://www.honeywell.com/sites/portal?smap=aero&page= aerotechmagazine3&theme=T4&catID=C4863EE40-DC5F-6161-7D42-3FDC5B5918B9&id= H35898141-2CF4-7CCC-08BF-59399596C9BF&sel=2.

Kramer, L., Prinzel, L., Arthur, J., & Bailey, R. (2004). Pathway design effects on synthetic vision head-up displays. *Proceedings of SPIE, 5424, 61-72.* doi:10.1117/12.543937.

NASA (2000). *Synthetic Vision Would Give Pilots Clear Skies All the Time.* Retrieved from March 3, 2009, from http://www.nasa.gov/centers/langley/news/factsheets/SynthVision.html.

NASA (2001). *Concept of Operations for Commercial and Business Aircraft Synthetic Vision Systems.* Retrieved March 3, 2009, from http://ntrs.nasa.gov/archive/nasa/casi.ntrs.nasa. gov/20020013800_2002002936.pdf.

Pope, S. (2006). *Cockpit Avionics: The Promise of Synthetic Vision—Turning Ideas into (Virtual) Reality.* Retrieved March 3, 2009, from http://www.ainonline.com/news/single-news-page/article/cockpit-avionics/.

Rockwell Collins. (2000). *Synthetic Vision Information System.* Retrieved March 1, 2009, from http://www .rockwellcollins.com/syntheticvision/docs/2000-DASC-SVIS.pdf.

Ruffner, J., Deaver, D., & Henry, D. (2003). Requirements analysis for an air traffic control tower surface surveillance enhanced vision system. *Proceedings of SPIE, 5081, 124–135.* doi:10.117/12.498997.

Scott, W. (2004). *NASA Team Bring Synthetic Vision to Maturity.* Retrieved March 3, 2009, from http://www.aviationweek.com/aw/generic/story_generic.jsp?channel=awst&id=news/08094top.xml.

Sindlinger, A., Meuter, M., Barraci, N., Klingauf, U., Schiefele, J., & Howland, D. (2006). Synthetic vision helicopter flights using high resolution LIDAR terrain data. *Proceedings of SPIE, 6226, 6226021–9.* doi:10.1117/12.664893.

Szoboszlay, Z., Jennings, C., & Tiana, C. (2006). Synthetic vision for rotorcraft: low level flight. *Proceedings of SPIE, 6226, 62260N1–12.* doi:10.1117/12.667891.

Vernaleken, C., Mihalic, L., Guttler, M., & Klingauf, U. (2006). A fresh look at runway incursions: onboard surface movement awareness and alerting system based on SVS. *Proceedings of SPIE, 6226, 6226J1–16.* doi:10.1117/12.664897.

Vernaleken, C., von Eckartsberg, A., Mihalic, L., Jirsch, M., Langer, B., & Klingauf, U. (2005). The European research project ISAWARE II: a more intuitive flight deck for future airliners. *Proceedings of SPIE, 5802, 13–24.* doi:10.1117/12.609862.

Watson, B. (2008). *Synthetic Vision: Seeing What Pilots Can't*. Retrieved March 3, 2009, from http://www.aviation.com/technology/080909-synthetic-vision-systems.html.

Wickens, C., Lee, J., Liu, Y., & Becker, S. (2004). *An Introduction to Human Factors Engineering* (2nd ed.). Upper Saddle River, NJ: Pearson Prentice Hall.

Williams, S., Arthur, J., Shelton, K., Prinzel, L., & Norman, R. (2008). Synthetic vision for lunar and planetary landing vehicles. *Proceedings of SPIE, 6957, 6957061–12*. doi:10.117/12.777079.

Yonemoto, N., Yamamoto, K., Yamada, K., Yasui, H., Tanaka, N., Migliaccio, C., Dauvignac, J., & Pichot, C. (2006). Performance of obstacle detection and collision warning system for civil helicopters. *Proceedings of SPIE, 6226, 6226081–8*. doi:10.1117/12.666693.

Airbus A380 Super Jumbo

Cefrey, Holly (2002). *Super Jumbo Jets*. New York: Rosen Publishing Group.

Fife, W, & McNerney, M. T. (1998). A Look into the Future of Airport Planning, Design, and Construction by Analyzing Current Issues. *In Airport Facilities: Innovations for the Next Century*, Retrieved March 12, 2009, from http://onlinepubs.trb.org/Onlinepubs/millennium/00002.pdf.

Lawrence, Philip (2006, September, 5). Growth, Capacity and Technology: Why the Airbus A380 will be a Major Commercial Success. *Morgan Stanley Research*, Retrieved March 02, 2009, from http://www.leeham.net/filelib/A380DEBATEFINAL.pdf.

McCormick, Carroll (2003, February). Getting Super Jumbos off the Ground. *Nickel*, Retrieved March 15, 2009, from http://www.nidi.org/index.cfm/ci_id/11065.htm#1.

Norris, G., & Wagner, M. (2005). *Airbus A380*. St. Paul: Zenith Press.

Proctor, Paul (1994). Super-Jumbos Pose Design Challenges. *Lexis Nexis Academic, 140*, Retrieved March 4, 2009, from http://lexisnexis.com.proxy2.li.uwo.ca:2048/us/lnacademic/frame.do?tokenKey=rs.

Singleton, A. (2008, August 12). *Airbus A380 - The Future of Aviation*. Retrieved March 23, 2009, from http://ezinearticles.com/?Airbus-A380—-The-Future-of-Aviation&id=1403233.

Sogame, Hiroshi (1993, July, 21). Lauda Air B767 Accident Report. *Aircraft Accident Investigation Committee Ministry of Transport and Communications Thailand*, Retrieved March 23, 2009, from http://www.rvs.uni-bielefeld.de/publications/Incidents/DOCS/ComAndRep/LaudaAir/LaudaRPT.html.

Spencer, Cam (2005). Airbus A380 Australian Airports Operational Issues. *Brisbane Airport Corporation*, Retrieved March 13, 2009, from http://www.asasi.org/papers/2003/A380%20Airport%20Considerations_Spencer.pdf.

Sutton, Oliver (1997). Super jumbos: new systems to boost performance, cut costs. *Interavia Business & Technology*, Retrieved March 02, 2009, from http://findarticles.com/p/articles/mi_hb3126/is_n611_v52/ai_n28690953.

TSB, (2006). Air Canada 143. Retrieved March 23, 2009, from Aviation Safety Network Web site: http://aviation-safety.net/database/record.php?id=19830723-0.

Blended Wing Body

"Air Travel Emissions Worse Than Predicted (Boeing 787 Airbus A380 airline emissions air travel)." Triple Pundit—Business, Better. Since 2005. 23 Mar. 2009 <http://www.triplepundit.com/pages/air-travel-emissions-worse.php>.

"BBC NEWS | Technology | 'Silent aircraft' How it works." BBC NEWS | News Front Page. 23 Mar. 2009 <http://news.bbc.co.uk/2/hi/technology/6120132.stm>.

"Blended Wing Body—New Concept in Passenger Aircraft." The Wing Is the Thing—Flying Wings and Tailless Aircraft. 23 Mar. 2009 <http://www.twitt.org/bldwing.htm>.

"Jet Fuel Price Monitor." International Air Transport Association. 23 Mar. 2009 <http://www.iata.org/whatwedo/economics/fuel_monitor/index.htm>.

NASA Langley Media and Public Outreach Offices. 23 Mar. 2009 <http://oea.larc.nasa.gov/PAIS/pdf/FS-1997-07-24-LaRC.pdf>.

"Quarterly Cost Index: U.S. Passenger Airlines." Air Transport Association. 23 Mar. 2009 <http://www.airlines.org/economics/finance/Cost+Index.htm>.

New and Emerging Markets

Zinnov, Research and Consulting, (2007). Global aviation markets—analysis. Retrieved March 21, 2009, Web site: http://www.zinnov.com/presentation/Global_Aviation-Markets—An_Analysis.pdf.

(2008, March, 17). Change in Canadian ATC taxi phraseology—line up/line up and wait. Retrieved March 9, 2009, from U.S. Department of transportation, federal Aviation Administration, Information for Operators Web site: http://www.faa.gov/other_visit/aviation_industry/airline_operators/airline_safety/info/all_infos/media/2008/info08013.pdf.

Fox, Kathleen (2009). AIP Canada (ICAO), Aeronautical Information Circular 15/05. Retrieved March, 1, 2009, from http://www.navcanada.ca/ContentDefinitionFiles/Publications/AeronauticalInfoProducts/AIP/Current/PDF/EN/part_5_aic/5aic_eng.pdf.

(2006). ICAO list of contracting states. Retrieved March 21, 2009, from International civil aviation organization Web site: http://www.eurocontrol.int/icard/gallery/content/public/icaoctry.pdf.

(2007). ICAO language proficiency requirements for pilots and air traffic controllers. Retrieved March 13, 2009, from English for Aviation Web site: http://www.englishforaviation.com/ICAO-requirements.php.

Webster, Ben (2008). Polish pilots' poor English almost led to crash. Retrieved March 13, 2009, from Time Online Web site: http://www.timesonline.co.uk/tol/travel/news/article4116523.ece

Gladwell, Malcom (2008). *Outliers, the story of success.* New York: Little, Brown and Company.

Fischer, Ute and, Orasanu, Judith (1999). Retrieved March 20, 2009, from Say it again, sam! effective communication strategies to mitigate pilot error Web site: http://www.lcc.gatech.edu/~fischer/ISAP99.pdf

(2006). Retrieved March 05, 2009, from ICAO's review of the international requirements for flight crew licensing and training Web site: http://74.125.95.132/search?q=cache:McyWXZtYNYwJ:www.jaa.nl/conference/19th/ICAO%2520PRESENTATION.ppt+icao%27s+review+of+the+international+requirements+for+flight+crew+licensing+and+training&cd=1&hl=en&ct=clnk&gl=ca.

(2007). Assessing compliance with ICAO standards and recommended practices (SARPs) and implementing their provisions. *U.S. department of transportation, federal aviation administration, national policy,* Retrieved March 15, 2009, from http://www.faa.gov/documentLibrary/media/Order/ND/1240.11.pdf.

Psutka, Kevin (2008, December, 12). Update on 406 ELTs - Our sector must equip. Retrieved March 16, 2009, from http://www.copanational.org/non-members/PresidentsColumn/2009/FEBRUARY/Update406ELTDec12.htm

Panorama of the European union. Retrieved March 17, 2009, from Europa Web site: http://europa.eu/abc/panorama/index_en.htm.

(1944). Convention on international civil aviation, signed at Chicago, on 7 December 1944 (Chicago Convention). Retrieved March 21, 2009, Web site: http://www.mcgill.ca/files/iasl/chicago1944a.pdf.

Future Impacts of Automation

Funk, K., Lyall, B., Wilson, J., Vint, R., Niewmczyk, M., Suroteguh, C., et al. (1999). Flight deck automation issues. In K. Funk, & B. Lyall (Eds.), The international journal of aviation psychology: Special edition: Aircraft automation (pp. 109–123). Mahwah, NJ: Lawrence Erlbaum Associates, Inc.

Harris, D. (2004). Flight instruments & automatic flight control systems (6th ed.). Oxford, UK: Blackwell Science Ltd.

Hettinger, L. J., & Haas, M. W. (2000). Current research in advanced cockpit display concepts. In L. J. Hettinger, & M. W. Haas (Eds.), The international journal of aviation psychology: Special edition: Current research in advanced cockpit display concepts (pp. 227–229). Mahwah, NJ: Lawrence Erlbaum Associates, Inc.

Hoeft, R. M., Kochan, J. A., & Jentsch, F. (2006). Automated systems in the cockpit: Is the autopilot, "george," a team member? In C. Bowers, E. Salas & F. Jentsch (Eds.), Creating high-tech teams (pp. 243–259). Washington DC: American Psychological Association.

Innovative future air transport system. (2007). Retrieved March 17, 2009, from http://www.ifats-project.org.proxy1.lib.uwo.ca:2048/.

Jukes, M. (2004). Aircraft display systems. Northgate, UK: Professional Engineering Publishing.

Kochan, A. (2005). Automation in the sky. 32(6), 468–471. Koop, M. (1998). 1920–1935 history–aviation resource centre. Retrieved March 16, 2009, from http://www.geocities.com/CapeCanaveral/4294/history/1920_1935.html.

Olsen, W. (2001). Identifying and mitigating the risks of cockpit automation. Retrieved from http://www.au.af.mil/au/aul/aupress/Wright_Flyers/Text/wf14.pdf.

Osder, S. S., & Hutchingson, I. N. (1961). In Sperry R.,Corporation (Ed.), Adaptive autopilot servo system. Delaware United States:

Theunissen, E. (2007). Increasing safety through better navigation displays. In H. M. Soekka (Ed.), Aviation safety (pp. 743–762). Netherlands: VSP.

Wells, A. (Ed.). (2001). Commercial aviation safety (3rd ed.). New York: McGraw-Hill.

Zuschlag, M. (2002). Human factors issues regarding automated flight systems. In E. Salas (Ed.), Advances in human performances and cognitive engineering research (1st ed., pp. 157–199). Amsterdam: JOI Press.